Quacks: Two years as a patient in a Veterans Administration nursing home

by **Fred Dungan**

Noun: quacks kwaks

1. Charlatans who pretend to have medical expertise.

Verb:

1. The harsh sounds of a duck.

2. To act or do business as a quack.

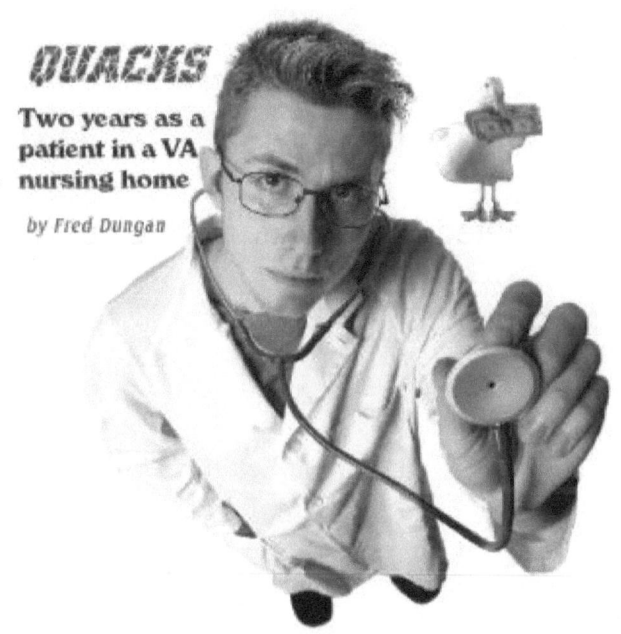

A DUNGAN BOOKS PUBLICATION

Published by DUNGAN BOOKS

First Printing in Paperback, May 2011

This is a non-fiction work about real people, the purpose of which is to present constructive criticism and incisive analysis so as to improve the overall quality of life for long-term Veterans Affairs nursing home inpatients

ISBN: 978-1-257-74971-3

3749 Myers Street

Riverside, CA 92503

(951) 688-1396

ISBN: 978-1-257-74971-3

Published and printed in the United States of America

fdungan@fdungan.com

Dedicated to the memory of my parents, Chief Russel Alonzo Dungan, U.S.N. (Retired) and Blanche Marie Dungan, whose selfless, God-fearing, caring natures were not lost on their son. Although you no longer walk this earth, your footsteps remain, suspended by good deeds in the sands of time. May your descendents continue to display these social attributes for innumerable generations to come.

TABLE OF CONTENTS

Prologue

There are not many books written about nursing homes. That is because nursing homes are insular institutions. Patients tend to be elderly and few write books. Far too many patients die there. Administrators blame the high death rate on old age. Old age is indeed a factor, but negligence and quackery take a high toll.

Few long term patients leave nursing homes alive. Dead people do not complain. Many people assume that their loved ones are being given adequate care and nutritious meals. Having lived in a VA nursing home for two years, I know better.

Nursing homes range from small, minimum care facilities where patients are largely warehoused on tranquilizers, social security checks have a way of quickly vanishing, and the nursing staff is underpaid and overworked to secure assisted living units where the patients live much the same way they did at home, activities are organized by experienced social directors, and it is possible to maintain privacy.

I would rate VA nursing homes as being in the top third of these facilities. Considering their bloated budget (which the taxpayers pay), there is room for improvement. Families should check out a nursing home before committing a relative to it. Some nursing homes are better than others. I am writing this book in the hope that it may help to improve the quality of veterans' healthcare in some small way.

Congress recently passed the Nursing Home Transparency and Improvement Act which requires nursing homes to disclose how good of a job they have been doing. Minimum quality standards will be put into place. Supposedly, negligence and abuse will be reported and a national registry will be established that will help nursing homes in identifying potential workers who have proved abusive and/or negligent at other institutions.

I am providing an example of proven quackery for veterans who may be under the opinion that VA healthcare does not have any overarching problems that result in unwarranted patient deaths. From May 14, 1999, to July 10, 2002, Paul Kornak was hired by the Stratton (New York) VA medical facility to direct research on sick veterans. One study involved injecting experimental cancer drugs into patients that caused a number of needless deaths. It was later discovered that Paul Kornak defrauded two drug firms out of $639,000. He pleaded guilty to the charges. He has been barred from federal employment for life. The VA failed to charge Paul Kornak with murder. Earlier, Kornak had been convicted of mail fraud. Why wasn't this discovered by a background check?

The following statement was posted by veteran and activist Sue Frasier on December 11, 2005, on an online media outlet called One Voice Veterans Forum:

"...We had gripping VA drama playing out here in Albany NY that was overshadowed by a sinister news media who

Paul Kornak, Stratton VAMC

did their best to confuse and conceal what was really a straight forward story about murdered veterans at a VA Hospital by a phoney...employee who was almost successfully protected by the VA itself.

Former VAMC employee Paul Kornak was given a lousy 6 year sentence for killing five.... that's right....5 veterans who were staying at the hospital, and this joke of a mandate was issued by another Army veteran himself, Judge Frederick Scullin. Scullin is notorious...for being the biggest pile of crap going when it comes to justice for our Veterans, so the fact that he gave Paul Kornak...a one year per murder sentence really did not come as any surprise to those of us who live here.

...Paul Kornak never finished medical school but somehow managed to get hired with fake medical credentials at the VAMC here in Albany, NY. He was placed in the Cancer Care unit and began signing his papers as Dr. Paul Kornak, and represented himself throughout the hospital in this same way. He began falsifying the paperwork of veterans staying in the cancer care clinic and entered the patients into drug treatment clinical trials which they were not eligible to be in. {As

Kornak's trial progressed, it also came out that he had falsified the patient records of 65 other veterans, but I digress.}

Two VA Pharmacists began to notice the lethal dosing that was being issued to the Veterans. They made attempts to intervene and were stopped. The 5 Veterans died....bingity, bangity, boom! The 2 Pharmacists went to the Administrators to report the lethal dosing and were promptly fired from their jobs. Yup -- that's right, they were FIRED !

They went to the FBI to tell their story, and wrote complaints to the VA Inspector General's Office...It was only after the pharmacists went out to the news media with other employees that the FBI finally agreed to review the case.

Paul Kornak was taken into federal custody on a 41 count grand jury indictment shortly after that.

The...VAMC Director, Mary Ellen Fache-Piche even had the stones to grandstand in a phoney press conference of sorts with the U.S. Attorney {talking about patient care and justice} to spin the entire story away from themselves. The news media fully cooperated with this pile of crap news conference even though it had been all over the area prior to that that the Administration fired the Pharmacists.

The two fired pharmacists were eventually hired back into their jobs, but only after a considerable lawyer expense to do so.

...Paul Kornak pleaded guilty to the indictment. In exchange for that, the government gave him a lousy one year per murder of a Veteran, a total of 71 months. He is in his 50's now, and he will be 60 or so when he is released.

The families had to take out second mortgages on their houses to pay for lawyer bills in a civil action against Kornak. The family lawyers used the Nuremburg...(ruling)...of the Nazi era as the main body of law for their pleadings. No word has ever been in the press about the results of the civil action.

Apparently, the...judge was on the same page with the orchestrated effort to diminish the importance of what had really happened.

Perhaps the biggest tragedy of all is that not one single veterans organization was at the string of hearings to give support to the family members. No DAV, no VFW, and no American Legion. On the day of the sentencing, one of the widows did the TV news interviews entirely alone by herself. There wasn't an organizational hat anywhere to be found next to her.

...Paul Kornak had an accomplice, a Dr. George Holland, who was fired from the VAMC at the same time that Paul Kornak was, but managed to flee the area and relocate somewhere in the state of Georgia, and is continuing to practice medicine...

If this story isn't your wake-up call that we are all in extreme danger inside of this system, then I don't know what is."

Considering Mr. Kornak's easily obtainable record of fraud, he could not have passed even a cursory scrutiny of his résumé. In 1993 in Harrisburg, Pa., Judge William W. Caldwell of United States District Court sentenced Mr. Kornak to a $2,500 fine and three years of probation for forging his credentials to obtain a medical license. Apparently, Mr. Kornak's history of fraud began with the falsification of a college transcript, and lie followed lie until he lost a medical license in Iowa, was denied one in New Jersey and was arrested in Pennsylvania.

Six years after Judge Caldwell's pronouncement, Mr. Kornak answered an advertisement for a research assistant position at the Albany veterans hospital's research institute.

Mr. Kornak told Dr. Hrushesky that he had lost his medical license because he could not document a year of medical school in Poland, according to the journal. Mr. Kornak "gave us a résumé with an M.D. on it and a lot of gaps," Dr. Hrushesky told the media. "We decided to give him a chance."

Unfortunately, the patients that Kornak used for guinea pigs in his macabre drug experimentation never stood a chance.

"Research violations were a way of life at Stratton for 10 years," said Jeffrey Fudin, a pharmacist at the hospital. "Stratton officials turned a blind eye to unethical cancer research practices and punished those who spoke out against them. The whole Kornak episode could have been prevented."

According to Paul Kornak's lawyer, E. Stewart Jones, there was a "clear systems failure," permitting a research culture where "rules weren't followed, protocols weren't applied and supervision was nonexistent."

It was also a culture whose descent into criminality forced the Department of Veterans Affairs nationwide to reckon with what an internal memorandum in 2003 described as "systemic weaknesses in the human research protections program, especially in studies funded by industry."

Excluding simple chart reviews, about 80 percent of the department's human research is financed by industry. The private sector pumps considerable cash into the system. In Albany, it accounted for $500,000 of the $1.15 million in research funding in 2004.

In January 2003, the VA abolished its independent research safety watchdog office, the Office for Research Compliance and Assessment (ORCA). The Bureau of National Affairs reported that "the move has puzzled Capitol Hill observers, some of whom have suggested an act of Congress to restore an independently functioning office." Obviously, despite being a public agency charged with transparency, the VA does not tolerate criticism and whistleblowers.

On March 25, 1999, Terence Monmaney, a medical writer for the Los Angeles Times wrote that a patient at the West Los Angeles Veterans Affairs Medical Center told doctors twice that he did not want to be a guinea pig. Nonetheless, they kept him on an operating table with an electrophysiology probe inserted in his heart for an extra 45 minutes to collect research data.

An inspection at a Veterans Affairs nursing home in Philadelphia in 2008 turned up conditions placing veterans at imminent risk of harm, including one patient whose leg had to be amputated after maggots were seen falling from his foot.

The multiple deficiencies at the facility, part of the VA Medical Center, were included in a 16-page report released to the Pittsburgh Tribune-Review in response to a federal Freedom of Information Act request.

The report by the Wisconsin-based Long Term Care Institute concluded that the facility, whose bed count has been cut from 240 to 120, "failed to provide a sanitary and safe environment for their residents."

"There was a significant failure to promote and protect their residents' rights to autonomy and to be treated with respect and dignity," the report concludes.

VA administrators, in response to the study, issued a corrective action plan, updated on June 29, which includes the hiring of consultants, additional staff, remedial training, and retraining programs for staff. "A great many changes have been instituted at the (nursing home) over the past year to improve the quality of care and quality of life for our veterans," VA spokesman Dale Warman wrote in an e-mail response to questions.

Citing "significant issues" with resident grooming, housekeeping and pest control, inspectors observed dried blood and tube feeding on the floors.

"Re-educate staff on dignity, respect and privacy," the action plan dictated.

Three months before the report was issued, a mute and disabled Vietnam veteran, David Allen, 56, died from choking on solid food, even though he was supposed to be on a soft-food diet. His death was not mentioned in the report, but triggered an internal investigation. In a written statement about the choking incident, the VA said the contracts of two agency nurses were terminated. Other staff members were given additional training on swallowing difficulties "as well as the effects of behavioral medications," according to the statement.

Allen's sister, Belinda Allen, said she was told her brother choked to death "and they did everything they could."

The records of the veteran whose leg was eventually amputated showed that no action was taken even though his toes turned black. After maggots were observed "falling out of the patient's foot," the amputation was ordered, the study said.

Besides that patient's case, the inspection report cites substandard treatment of wound care and "multiple concerns regarding nursing competencies."

In a case witnessed by an inspector, a nurse applied the wrong medication to a wound despite a week-old order from a doctor changing the prescription. Inspectors found serious deficiencies in the care of veterans dependent on tube feeding, including a lack of documentation that tubes were flushed as required, or that proper sanitary practices were followed.

Some of the patients experienced substantial weight loss, including one veteran with an unexplained 51-pound weight loss.

"The potential for dehydration for these residents presents immediate jeopardy," the report says.

The report criticizes facility staff, citing complaints from patients and their own observations. One resident, asked to explain a wound on his arm, told the inspector a nurse "just ripped the dressing off. She just ripped it off real quick and all my skin came with it," the report states.

In Marion, II, care of veterans and hiring practices have become a topic of discussion after an investigation into the VA Medical Center was made public. The investigation concerns Dr. Jose Veizaga-Mendez, who is allegedly responsible for more than a few surgical deaths at the center. The physician was hired by the VA despite being forced to give up his license in Massachusetts because of multiple malpractice cases. All inpatient surgeries at the Marion VA center were suspended on August 31, after administrators noted an increase in the hospital's mortality rate. From October 2006 to March 2007, nine people died after surgeries involving Veizaga-Mendez. Two deaths are generally regarded as normal for the time frame.

The October 28, 2007, edition of the Boston Globe highlights the story of Robert A. Whitney, who suffered in pain for almost four years after his hernia operation. It turns out that Veizaga-Mendez embedded surgical staples into Whitney's bladder. In fact, Massachusetts officials allege that Veizaga-Mendez provided dangerous care to at least seven other patients, two of which died as a result of the care. This all occurred before Veizaga-Mendez was hired by the Marion VA Medical Center.

One of the patients at the Marion VA died from internal bleeding after gallbladder surgery. The victim's wife has filed a notice of intent to sue the VA, alleging that the VA should not have let Veizaga-Mendez perform surgery at all.

That a doctor who is under investigation in one state can be hired by the VA in another state is a cause of concern for VA bureaucrats. Lawmakers are criticizing the VA for not getting enough information about Veizaga-Mendez before hiring him. Four hospital officials have been reassigned until an investigation into the situation is completed.

Meanwhile, a United States army veteran has filed a lawsuit against an Albuquerque VA healthcare center after he suffered through a month of serious pain caused by allegedly inept medical staff. According to the veteran, mistakes made by medical staff include failure to remove surgical hardware from his leg even though it was infected with staph, refusing to allow him to seek medical treatment elsewhere and puncturing a vein while inserting intravenous lines, which allowed a blood clot to form and antibiotics to seep into his soft tissue.

In November 2008, the VA Medical Center in Augusta, Georgia sent a letter to more than 1,200 patients who were treated for ear, nose and throat, warning them they may have been exposed to infections.

After a December 2008 investigation at the VA clinic in Murfreesboro, Tennessee, officials discovered that healthcare workers were not properly maintaining the medical equipment used to conduct colonoscopies. More than 6,000 patients were notified and offered free testing.

In March 2009 VA officials announced that veterans in South Florida may have been exposed to Hepatitis and HIV after being examined with contaminated medical equipment.

According to reports, more than 3,200 veterans who received colonoscopies at the Miami VA medical clinic between May 2004 and March 12, 2009 are at risk of exposure to both Hepatitis and HIV.

Mr. Terry Soles, a Vietnam veteran, went to a VA hospital in 1998 complaining of diarrhea and pain where his doctors removed small, cancerous growths from his stomach and esophagus. The VA hospital subsequently administered many painful tests, but lost the results. After months of misdiagnosis and failure of treatment, Mr. Soles' cancer spread rapidly and killed him three days later.

In late December 2007, U.S. Representative Ginny Brown-Waite grilled a panel of doctors and administrators from James A. Haley VA Medical Center in Tampa, Florida concerning how closely they are supervising unlicensed psychologists who treat veterans.

At the same time, she questioned the motivations of a staff psychologist who filed a complaint with the state saying he is concerned that veterans unknowingly are being treated by unlicensed psychologists who are inexperienced and not getting enough supervision.

Michael Swango, VAMC killer

It was that complaint, from Brian Nussbaum, that brought public attention to the issue, resulting in Representative Brown-Waite convening the forum at Haley after she read newspaper reports about it.

Congresswoman Brown-Waite, Republican-Brooksville, pointed out that Nussbaum was not granted the title of team leader of the Post-Traumatic Stress Disorder clinic when he applied for it.

"Perhaps Dr. Nussbaum felt excluded," she commented after the forum, and added there appears to be a "breakdown in communication" within the program at Haley.

When asked by a reporter whether she has concerns that veterans are not getting top quality mental health care, she said the situation still should be monitored.

She asked most of the questions during the forum, as she sat alongside the Haley doctors and administrators. U.S. Rep. Gus Bilirakis, Republican-Palm Harbor also posed a couple of questions.

About 70 people attended, including leaders of veterans' service organizations. They were not given a chance to query the Haley staff.

The forum put Haley administrators on notice their program is being watched, said Verlin "Buck" Rogers, founder and past president of a local chapter of Korean War Veterans Association.

"I think we opened the eyes of the administrators, and they finally realize they need to listen to the veterans more," Rogers declared.

Rogers' chapter had voiced concerns that two longtime staff members, including one who had helped develop the PTSD program, no longer are working there. When they pressed for answers, Rogers said the hospital gave them "lip service."

Michael Swango graduated from the Southern Illinois University Medical School in 1983 and began the internship program at Ohio State University Hospital upon his graduation. While working as an intern at Ohio State University Hospital in January 1984, Dr. Swango murdered Cynthia McGee by injecting her with a lethal dose of potassium. In February 1984, he assaulted his patient, Rena Cooper, by injecting her with a poisonous substance. She survived the attack. After that assault, Ohio State University Hospital removed Dr. Swango from the residency program, and in 1985 Ohio authorities commenced a murder investigation into his activities. Although that investigation did not result in the filing of charges against Swango, he did learn of the investigation and concealed the fact that he was investigated for murdering patients from the other hospitals that subsequently hired him.

Swango served five years in an Illinois prison for aggravated assault stemming from injecting patients with poisons. Swango sought admission to several medical residency programs. In 1992, he was hired by the University of South Dakota and assigned to work as a resident at the VAMC Sioux Falls, South Dakota, after he falsified facts about his prior criminal conviction. Swango was discharged from the program after hospital administrators became aware of the facts surrounding his conviction and his activities at Ohio State University Hospital. In 1993, Swango applied for and obtained a position as a medical resident at the State University of Stony Brook Medical School, which ran a residency program at VAMC Northport. During the application process, he misrepresented that his criminal conviction in Illinois stemmed from a barroom brawl; a false statement that ultimately led to his conviction and incarceration on Federal charges. Thereafter, Swango murdered George Siano, Aldo Serini and Thomas Sammarco, while all three were patients at VAMC Northport. Swango killed all three patients by administering injections of toxic substances. In addition, Swango also injected a poison into another patient at the hospital, Barron Harris. Mr. Harris survived the incident. In October 1993 Swango was discharged from his residency at VAMC Northport.

In June 2000, Swango was indicted for three counts of

murder, one count of assault and one count each of false statements, mail fraud and conspiracy to commit wire fraud. Three years later, he was finally tried for the murders he had committed. On July 11, 2000, Swango pleaded guilty to killing three of his patients, and to fraud charges. He was subsequently sentenced to life imprisonment without the possibility of parole. Although Swango was ultimately convicted of three murders, it is estimated that he committed somewhere between 30 and 60 murders throughout the course of his career.

The warning signs should have been obvious. How could the VA have failed to detect his unfitness for duty during two separate hiring investigations? A residency program abruptly curtailed. A criminal conviction that led to prison time. A previous employer reluctant to provide background information because of fear of being sued. But the Veterans Affairs Medical Center in Sioux Falls, Iowa, in July 1992, and a year later at Stony Brook Health Sciences Center in Northport, NY, both welcomed Michael Swango to residency programs. Had they checked, either hospital would have known Swango was accused of killing several patients, and trying to kill others by injecting them with poison and as a paramedic, he had been found guilty of lacing coworkers donuts with arsenic.

Clearly, Veterans Affairs regularly hires unlicensed physicians—quacks—who pose a danger to their patients. Also, veterans are routinely used as guinea pigs for drug studies. The doctors are in league with the pharmaceutical companies that handsomely subsidize experimentation on veterans. Victims often are not told that they are subjects in a clinical study. When quacks are uncovered, the VA rarely seeks the maximum sentence for their crimes. I want to see these deviants prosecuted to the fullest extent possible under law. Our security and safety demands it.

Preface

Dr. Reipzig slid the x-ray of my right knee out of the large manila envelope and slapped it onto the backlit screen that hung on the examination room wall. Scowling at the negative, he pronounced the verdict: "kaput," which pretty much summed up its condition. Bone on bone, no cartilage left, it was time for me to get an operation.

Artificial knees usually last twenty years or more. When they wear out, they are relatively easy to replace. How do I know? It says so in the pamphlet the VA gave me to read.

The clerk at the front desk scheduled a pre-op appointment for me with the VA hospital's orthopedic surgeon. I drove home, wrote the pertinent information on a wall calendar, and gave the matter no more thought. Three months later, I received a phone call from a robot phone calling machine reminding me of it.

Chapter 1

Pre-op

After taking my temperature and blood pressure, the nurse directed me to one of the 50 or so identical examination rooms in 3NW Ortho and told me to sit down on a stainless steel piano stool. I waited for what seemed like eternity and it crossed my mind that they might have forgotten about me. I was about to stand up when the door opened and in stepped a flaxen haired young fellow in khaki trousers and a polo shirt. He looked as if he belonged on the 18th green. All that was missing was his titanium putter. I thought a white lab coat and stethoscope were de rigueur for Veterans Administration doctors. But this wasn't just any old doctor. Surgeons tended to be prima donnas. They comprised the top echelon of the VA medical hierarchy and were pretty much free to say and do what they pleased.

"Hi, I'm Dr. Harold Gustafson. I'll be leading the surgical team that is going to replace your—he paused to scroll up and down a file on a flat screen monitor—right knee."

The creamy white right hand that he extended to me was frighteningly pale—as if it was regularly kept tucked away and had only been brought out now because manly tradition demanded it. Obviously, it was a hand adverse to physical labor—no calluses here—with carefully manicured nails and cuticles. For a nanosecond I searched in vain for a trace of clear nail polish, but none was to be found. At least he's not gay, I thought.

My grizzled right paw took hold of his cold, clammy hand and pumped it up and down as if to wring out the excess moisture. No question about it, this guy had the grasp of a dead fish and I was sorely tempted to squeeze some life back into it when it occurred to me that this was the hand of a skilled surgeon who would shortly be cutting apart my knee with a glorified saber saw. Best to keep on his good side.

Then I noticed the Rolex on his wrist. It too was limp. Strange, why would someone buy an expensive watch and not have the metal band adjusted? Perhaps it wasn't his watch after all. Or maybe he had lifted it from a corpse after an unsuccessful operation. It took me back 30 years to a distant jungle where I used my bayonet to mine gold nuggets from the teeth of the enemy dead. It certainly wasn't anything I was proud of. But my sergeant was doing it and I desperately needed the money to get drunk enough to forget what I had become—a predator who was loathe to let morals and ethics interfere with the instinct to survive. Having pumped the limp appendage dry, I released it, half-expecting it to fall off the body and flop around on the floor. But then those kind of things almost never happen in real life. Kind of a shame, isn't it?

But I digress. My right knee was a real mess. For 10 years I had been walking stiff-legged with a knee that would not bend. The cartilage had worn away and calcification had fused what remained of the knee cap to the upper and lower leg bones. Bone on bone is about as painful as it gets—stabbing pain that wouldn't go away even when I wasn't walking. I was getting two to three hours sleep at night and popping far too many pills.

"So, how do you go about giving me a stainless steel knee," I asked.

"It's titanium," he corrected. "We use titanium because it is stronger than steel and weighs less than aluminum. They normally cost $50,000, but the VA gets a good price on them by purchasing in bulk. Your new prosthetic will last 20 years or more. And it will be easy to replace because it's crenellated."

"What's 'crenellated?'" I interrupted.

"It's like the battlements atop a castle wall. We will notch your leg bones to mesh snugly with the ridges of the prosthetic, much like the gears in a transmission. After a while, you won't even notice that it's there."

Dr. Gustafson smiled a bit too smugly for my tastes. Did I really want to put my life in his hands? The answer was a resounding no, but I couldn't come up with any other option. A private surgeon would cost more than I could afford. Besides, the VA was going to do it for free plus pay me a monthly check until I completely recovered. What could go wrong? Knee replacement was commonplace. It was so safe that the VA didn't hesitate to replace the bum knees of World War II vets who weren't in the best of health. The government was offering me an outright gift and it didn't seem right to dig too deeply into the details. I didn't want to appear ungrateful, however, there were a few questions for which I lacked answers. Seeing as how there is no time like the present . . .

"What's your rate of success? I mean, does anything ever go wrong?"

His eyes narrowed and he wet his lips. Unwittingly, I seemed to have struck a nerve, but it only took a moment for the smug smile to return.

"My team has the best statistics at this facility or any other VA hospital in this region," he bragged. "More than 99 out of 100 knee and hip surgeries are successful. I'm very good, in fact, I'm the best."

That was exactly what I wanted to hear. In my youth I had been a boxer. Statistics made sense to me. You are either a winner or a loser. I had seen enough winners to know that they oozed confidence from every pore. If you're going under the knife, the last thing you want is to have a loser perform the operation. I nodded my head up and down in approval.

Dr. Gustafson seemed relieved. He checked his surgery schedule and tentatively penciled me in for the first of three operations he would be doing on a Thursday three weeks from now. His assistant would provide the details. Like a fool, I signed a VA form authorizing surgery without bothering to read it. One more perfunctory handshake and I went out the door. Total face time with the man who would be sawing me apart had been less than five minutes.

Instead of shoehorning myself into a crowded elevator, I went down the stairs to the main lobby and exited the hospital through a side door. On my way back to my truck, I stopped to gather my thoughts at the manmade duck pond which encircles the facility in much the same manner as a moat encircles a castle.

This was elective surgery. Nobody was putting a gun to

my head. Other options were available to me. Loma Linda Medical wasn't the only VA hospital in Southern California. I could have said "no" and went elsewhere. There were VA hospitals in Long Beach, Los Angeles, and San Diego—all of which had skilled orthopedic surgical teams that had put artificial knees and hips in thousands of veterans. That was exactly what bothered me the most. No matter where you went in the VA system you ended up getting factory-style surgery. Gone was the close relationship between doctor and patient that was standard in private practice. I really didn't trust that limp hand to wield a scalpel.

So why had I signed the authorization form? As I was watching a pair of snowy egrets stand stiff-legged at the edge of the pond, the answer came to me. It was a piece of philosophy that had kept me sane when Uncle Sam snatched my conscientious objecting teenaged ass out of the University of California at Irvine and thrust me into the role of a shotgun toting, .45 caliber semi-automatic pistol equipped United States Army Military Policeman some 35 years earlier: "Oh well, what the hell?"

Chapter 2

Op

Just because I signed the authorization form, did not necessarily mean I was going to have an operation to remove my frozen right knee and replace it with an artificial factory-built gizmo that was guaranteed by Johnson & Johnson (the baby wipes people?) to last twenty years. But what if it broke down before that? Was I supposed to unzip my skin, rip out the bloody thing and send it via Federal Express overnight delivery to their plant in Dearborn, Michigan (or wherever—if you have seen one dilapidated Rust Belt city, you've seen them all).

Before the VA would pay the $50,000 plus surgical cost, I would have to prove that I was worth that much. After all, if I dropped dead during surgery, the government would lose a bundle of money. I mean, no matter how good the guarantee was, one could hardly expect a manufacturer to restock an artificial knee that had been salvaged from my decomposing corpse. Not that my corpse would be any different from any other veteran's corpse in the identical graves that flow row upon row like some horrendous crop that war has grown in our national cemeteries. No, it is simply that once you drive it off the lot, that knee is only worth a fraction of its original value. Think of it as a new car after it has lost that new car smell.

Everybody got a piece of me. There were EKG's to

perform, blood tests to be taken, urine to be sampled, and countless forms to fill out in triplicate (had Congress met in secret and repealed the Paperwork Reduction Act while we weren't looking?). Somewhere there is a set of rules and regulations that govern the VA. Veterans aren't permitted to see it, but we are nevertheless expected to abide by it. We refer to it as the VA Bible. In Genesis, it says that "an appointment begets another appointment which begets yet another appointment" and so on until there are enough appointments to give everyone in its bloated bureaucracy full employment with no danger of layoffs. And, should the budget shrink, benefits to the veterans are cut while medical personnel continue to enjoy their Garden-of-Eden lifestyle. Veterans function as guinea pigs unnecessarily sent to specialist after specialist with little or nothing ever coming from it. The primary purpose of an appointment is to schedule the next appointment. Frequently the physician is overloaded with patients to the point where there is hardly time for anything else. On busy days doctors are rushed to the point where they diagnose patients with diseases and ailments they do not have and for which they are prescribed drugs they would probably be better off not taking. Veterans are regularly kept busy running in circles with little or nothing to show for it. Most instinctively know better than to attempt to change the VA bureaucracy's attitude that they are dispensing a charity, rather than providing an entitlement. Besides, an overworked physician soon learns that it is easier to prescribe "feel good" pills than to determine the actual source of an ailment. It should not surprise anyone that even the most idealistic physicians eventually succumb to the regimen of a system of public healthcare that is substandard to that of private practice. This phenomenon is by no means limited to the Veterans Administration. Even in Honoré de Balzac's day, government employees who valued their jobs learned to go with the flow and got along to get along. The primary difference between then and now is one of degree in that the more money that the public pumps into the VA in an effort to help injured vets, the more bloated the VA becomes. Top VA administrators want nothing more than to add to the size of their fiefdom in order to aggrandize their personal standing. The objective is to further the image of the VA and thereby perpetuate the bureaucracy. All else, including healthcare is secondary. Thus, the construction and maintenance of buildings and grounds receives priority over healthcare for veterans. This can assume ridiculous proportions, e.g. at the Jerry L. Pettis Memorial VA Hospital in Loma Linda, California, disabled vets bide their time waiting for a walker or a wheelchair while duck ponds, valet parking, and state-of-the-art, motion-sensing self-flushing toilets receive priority. Or as the veterans say, it's all show and no go, with the emphasis placed on what VIP's see when they tour the place.

The ideal situation would be to have the VA run by veterans for veterans. If it was staffed with veterans, the agency would be sensitive to the needs of veterans. Most of the medical personnel were recruited from overseas countries such as South Korea and the Philippines ostensibly because immigrants will work larger caseloads for lower wages. No matter that many of them have a hard time learning English and/or tend to stick to themselves—to the detriment of veterans. There is nothing in this world as rude and disheartening as going to the VA clinic for a checkup and having a couple of Filipino nurses discuss your condition in their native language, Tagalog, in front of you. Why are they being so secretive? Could it be that they know something that you don't—like maybe you have leprosy or AIDS? It's just plain bad manners to talk about someone who is present in the room as if they are too dumb to understand what you are saying. If they must hire their employees from the dregs of foreign cultures, the Veterans Administration should at the very least give them an idea of how they are expected to act. There is a lot more to being a good nurse than processing the maximum number of patients in the shortest time.

Forgive me for going off on a tangent, but I'm an old man and I tend to ramble. That's the only nice thing about getting old—people make allowances for your age, so I can get away with quirky behavior. When you are young and poor and you do outlandish things, people call you crazy. But as you get older and acquire some wealth, they either shut their mouth or refer to you as eccentric. You are the same person behaving just as badly as before, but somehow your age and money makes a big difference.

As I said earlier, I had to have my blood tested and my urine analyzed. That takes place at the Laboratory where you have to take a number and wait. A cylindrical device on a stand spit a scrap of paper with the numbers 069 at me. There was row after row of government issue dingy orange fiberglass bucket chairs in the waiting room and each chair was bolted to the others in the row by means of a heavy metal bar. Evidently they were afraid that the veterans were going to steal their ugly beat-up chairs. I tried to picture myself going down the stairs with a chair under each arm and a VA police officer closing in on my tail while shouting "halt or I'll shoot," but it was overly melodramatic and ridiculous; no such thing had ever taken place and none ever would. On the far wall an electronic board flashed the number 003. That meant there were 66 people ahead of me. Sixty-six people waiting to be checked in by one of the clerks at the front counter. And after that I would have to wait to have my blood drawn by a technician. Long waits are commonplace at the VA. I might as well get used to it because it is the only healthcare I have. Anything is better than nothing.

After waiting for what seemed like forever, my number came up. Oh, happy day! There is a God and he has heard my plea and delivered me from suffering the humiliation and pain of this accursed ergonometric bucket seat. My butt had gone numb and my back throbbed. I was stiff all over. Because my right knee could not move, I had to rock back and forth to stand up. Meanwhile, the clerk at the counter had seen nobody coming and had gone on to number 070. A beer-belly fellow in a white "I LOVE NY" t-shirt had handed over number 070 just as I got to the counter. Brazenly reaching around in front of him, I placed number 069 on top of the pile of numbers on the counter and said "I believe I come first." I realize now that Miss Manners might not have endorsed my boarding house reach, but I had been waiting for almost four hours and I wasn't about to start over again. I had pushed Number 070 and was about to find out what happened when you shoved a plug-ugly Marine out of the way. He wasn't about to take crap from the likes of me and said it so loudly that a VA policeman was summoned. The officer stood between us as a phlebotomist drew a tube of blood from each of us and he subsequently escorted us to the parking lot. He had given us a break, but somehow neither of us was feeling grateful. Nobody should have to wait four hours to have their blood drawn. There is nothing like wasting a half day in line at the VA to bring out the righteous indignation in a well-meaning, God-fearing veteran. Standing in line waiting for the government to begrudgingly give you your due is a socialist disorder that has no place in the United States. In the Soviet Union people had to stand in line for a roll of toilet paper. We won the Cold War, didn't we? Then why are our veterans standing in line? I see no reason why the Veterans Administration cannot function in an organized and efficient manner. One blood test that took less than five minutes to perform was made to kill an entire day of my time. If such an inefficient rate is deemed desirable, why not go back to making house calls? The technician's time is valuable, but so is everybody else's.

The following morning I returned to the hospital for an

electrocardiogram. This time there was no waiting. My dear departed mother couldn't have treated me better. I had no sooner checked in then the cardiology technician called my name. I was done in a few heartbeats. Afterwards, she had me hand carry the test results to the surgeon's office to make sure the paperwork couldn't get lost in the interoffice mail. Why couldn't the Lab be like this? It's not that difficult to do the job right. Was it a morale problem? The clerk at the Lab simply went through the motions without even bothering to look up. 069, 070, it was all the same to him. Three hours to Miller time, two more days till payday. What happened to the poor unfortunate patients wasn't his concern. After all, things were looking up. Only two hours and fifty minutes to Miller time. He could close his eyes and feel the cold beer massaging his throat as it went down. Two hours and forty-seven minutes to go. God willing, he will choke on his beer and the next time I come to the VA hospital for a blood test, the slacker that used to work at the counter will be toasting his tootsies in Hell.

The next stop was the Radiology department where I was scheduled to get some x-rays of my right knee. They had me strip and put on a gown. Then they wanted me to climb up on a cold slippery stainless steel table that stood three feet off the floor. Ever try to jump on a table without bending your knee? It simply cannot be done. Besides which I am flatfooted and don't have an arch from which to spring. Nevertheless, the radiologist encouraged me to do it. Of course, the two people on duty could have helped me up, but then where is the fun in that? Much better to comment on how I didn't seem to be trying hard enough. They were being paid by the hour and could afford to wait all day if necessary. Eventually, they tired of their sport and took the x-rays with an older model (old enough to have been used by Madame Curie) vertical axis machine. They weren't happy with the results, but hustled me off anyway. Having an x-ray done with black and white film is about as close as I will ever get to film noir. Did James Cagney have a bad knee or was it George Raft? Call me a gimp, but I'm in good company.

I don't mean to belittle x-rays. Far from it. If it weren't for x-rays, doctors would be in the dark. However, it wasn't so very long ago that children's shoe stores used x-ray machines to convince mothers that their child had been properly fitted with shoes. You put your feet under a vertical box while your mother observed an x-ray image in bright green and black through a metal stereoscope on the top of the box. Lord only knows how much dosage we got.

The runaround is almost over. All of the specialists on the checklist that Orthopedics gave me have examined me and declared me fit for surgery. I guess that means I'm 100 percent USDA—or VA—Grade A choice beef. Oh happy day! I'm good enough to butcher. Surely, I am truly blessed.

A week later, the mail carrier delivers a letter from Ortho informing me that I am going to undergo surgery. I will be the second of three veterans who will be fitted with an artificial knee on the appointed day. I should come an hour early and deposit my valuables with the clerk. No eating after midnight the night before the operation. Nothing to drink on the day of the operation. No smoking. However, nary a word about sex. Although the instructions didn't say it, I gathered that the thing to do was to shack up the night before so that if they slipped up and slashed me to death with a scalpel, I would at least die happy. Be prepared for the worst case scenario. That way you will be pleasantly surprised when you don't end up on a cold slab of stone with a tag on your big toe and embalming fluid in your veins.

The medical profession would like for us to believe that operations are routine; there is no need to worry. Make no mistake about it, being cut open is always a big deal. If you don't die on the table, there is still a risk of infection. There are no guarantees. Getting a second opinion should be mandatory. For all you know, you might be getting an operation so that the surgeon can make the next payment on his boat. Never trust someone who wears deck shoes to cut on you.

Why am I so wary? A year ago quackery (the unrestrained business of medicine) had been responsible for murdering my father, a retired United States Navy quartermaster, who had survived numerous World War II landings in North Africa and Europe, only to be brought down by a two-faced heart specialist who made more money by treating him with pills for ten years than he would have if he had referred him to a surgeon to get the heart bypass he needed. In the end, my father, who was covered by three different health plans (Champus, Pacific Care, and Medicare) was sent to a county hospital to die because the HMO's were being run by administrators who argued over how much of the bill each should pay. After the funeral, I went to Pacific Care's offices to complain. They could have cared less. I had to pound my fist into his keyboard to get the administrator's attention. Accountants aren't qualified to make medical decisions. Why wasn't there a physician in charge? Profits take priority over patients' lives. No wonder we are plagued with assisted suicide and late term abortion. When religion, morals, and ethics are brushed aside, there is nothing left to hold society together. Our forebears had the courage of their convictions. How far must we fall before we are willing to take a stand? Nobody gives anyone anything. The price of liberty is eternal vigilance.

Nor am I ungrateful for the services that taxpayers are providing me. I am simply mindful that bureaucracies often have hidden agendas. This is not a modern phenomenon, nor is it peculiar to our form of government. In the early 19th century, Honoré de Balzac wrote that in dealing with bureaucracies "you cannot prevent the buying and selling of influence, the collusions of self-interest." Has my surgeon accepted gifts from a prosthesis manufacturer or drug company? Surveys show that healthcare professionals who don't are an exception rather than the rule. Corruption is rampant and pervasive. Don't be naïve; your chances of surviving surgery could be influenced by such factors. The Hippocratic Oath is not what it used to be. When the American Medical Association conducted an in-house investigation, they found that of the medical schools that were still administering some form of the oath to their graduates, only 43 percent had them vow to be accountable for their actions. How pathetic! Moreover, a mere three percent prohibited sexual contact with patients.

We pay more for healthcare than anyone else. In 2006 the United States paid 15.7 percent of its Gross National Product to the medical industry, which is over five times the amount we spent in 1950. In the last eight years alone, insurance rates have doubled. Something is definitely amiss.

The day I am to have knee replacement surgery finally arrives. There is nothing left but to do it. Before I leave, I say goodbye to my faithful dog, Speedo. Although neither of us are aware of it, we will not see each other again for two years. I will miss a lot of things, but I will not miss anything as much as I miss him.

I walked into the VA hospital under my own power. That is important because I came out in a wheelchair two years later, worse off than when I went in. The Hippocratic Oath says "above all, do no harm." Society has a right to protect itself; no oath, no license. I have no use for abortion and assisted suicide. Margaret Mead wrote that "throughout the primitive world, the doctor and the sorcerer tended to be the same person." Surely, we have evolved beyond that. We need to rid ourselves of the rotten apples who give medicine a bad name.

We cannot expect doctors to be perfect. Anyone can make a mistake, but mistakes have to be acknowledged and corrected. Otherwise, they will no doubt be repeated.

The nurse asked me to undress and put on a hospital gown. It had evidently been designed for double-jointed patients because it tied in the back. Besides which it was far too flimsy to provide any protection from the cold. Breathing cold, stale, processed air makes people sick. They need to turn off the air conditioning, open the windows, and get some fresh air in here.

The two orderlies who are wheeling me on a gurney towards the operating room think I'm joking. Goosebumps aren't funny. If the sheet wasn't tucked in all the way around me like a mummy, I would scream "Airborne" and hit the ground running (or stumbling in a rapid manner).

The double doors of the operating room slam shut behind me, signifying that there can be no turning back. Ahead lies the brightly lit faces of the sardonically smiling surgical team.

Stainless steel instruments are being arranged on stainless steel trays. The surgeon takes a ballpoint pen from his pocket, clicks it open, and draws a line down the center of my right knee. A nurse asks me to remove my dental bridge and eyeglasses. I hand them to her and she puts them on a tray. Down comes a rubber mask that covers my mouth and nose. I am instructed to breathe deeply and count backwards from ten. Ten, nine, eight....

Chapter 3

ICU

The anesthesia wore off slowly. I was drifting in and out of consciousness. When I finally opened my eyes, I found myself in a white windowless room with tubes and wires coming out of my body, feeling terribly alone.

But I wasn't alone. There was a rather muscular fellow standing nearby who was wearing a wide black elastic contoured back support belt like those worn on the job by stevedores, warehousemen, and other workers who routinely do heavy lifting. How strange, I thought, what's there to lift in here?

The mystery man turned out to be Stan, a physical therapist assigned the mission of being there when a patient first woke up following surgery.

"Get up," he ordered after introducing himself.

"No can do," I said, gesticulating towards the tubes and wires attached to my body.

"You either get up now or your knee will freeze up," Stan asserted with an air of authority that set off a drill-sergeant alarm in my head. "Take a couple of steps and you can go back to bed."

Could I trust him to keep his word? I sincerely doubted it.

Nonetheless, it seemed futile to resist.

It took me a while to inch to the edge of the bed. My new knee felt like it was on fire. The pain was unbearable. I tried and failed to stand up. More than anything else, I wanted Stan to go away. However, it was his job to start me walking. He suggested I transfer to a chair by sliding down a wide plank he referred to as a transfer board. Since the slope was steep, it should have worked. Only it didn't. Perhaps I can't slide because I don't weigh enough, or maybe my skinny butt acts like a suction cup. Who knows? I sat at the top of the board and probably would be there still if Stan had not grabbed me in one hand (like King Kong did to Faye Raye) and dropped me on the seat. One thing is for sure. I now know what the belt is for.

It took awhile, but Stan eventually tired of bullying me and wandered off to torment some other poor veteran. I thought I was rid of him, but Stan returned fifteen minutes later with a vigor that told me he had made short work of his victim. Fortunately for me, Stan's shift was nearly over. I imagined him hurrying home to beat his wife and kick the dog. Perhaps that didn't happen, but I wouldn't be surprised if it did.

Later that night, I cajoled a nurse's aide into getting me a PVC potty chair on wheels from an empty room down the hall where she had seen it on her rounds. Anything is better than a cold metal bedpan. She helped me to get down off the bed and walk a couple of steps to where she had it parked. Somehow a few encouraging words from her accomplished a lot more than a litany of unspoken expletives from Stan.

Despite spending more than two years as an inpatient at the VA hospital, I never saw Stan again. But I often heard about him. Since there isn't all that much an inmate (excuse me, I mean inpatient) can do (especially when confined to bed), gossip and rumors are rampant. Although there were more than a dozen physical therapists, none of them even came close to matching the chutzpah of Stan. But I'm not worried. Guys like Stan usually end up as a notch on someone else's belt.

The Intensive Care Unit (ICU) resembles solitary confinement. I was in a room by myself tethered to a glucose intravenous drip and several monitors. No doubt it was necessary for the first few days, but I ended up stayed in ICU for three weeks solely because the nursing home on the first floor was full and, according to them, they had no other place to put me.

I used the phone on the nightstand beside my bed to make a few inquiries. What they said was true, but they didn't bother to say why. It was painfully clear that more veterans were admitted than discharged. The surgeons were performing more knee and hip replacements than the recovery and rehabilitation system could absorb. Unlike private hospitals, the VA hospitals preferred to do this with inpatients (perhaps because much of their rehabilitation resulted from war wounds). The process was lengthy and effective, albeit extremely slow. Pack too many patients into the wards and the risk of spreading disease and infection goes up. But the surgeons weren't concerned. Surgeons cut and stitch and call it a day. Dealing with the aftereffects isn't what they like to do. In a private hospital the doctor who admits you closely monitors your progress throughout your stay. Perhaps the VA hospitals could learn something from them. Nobody wants assembly line surgery. Chickens are butchered in that manner. I'm not meat. I demand to be treated with dignity. Whether a human being lives or dies should be a matter of importance to everyone, including the surgeon. Quacks don't care what happens to their patients. No matter how skilled the surgeon, he's a quack if his sole reason for doing his job is the money.

During my three week stay in the intensive care unit, my intravenous drip was being laced with morphine by a machine that injected the drug whenever I pressed a button. However, it would only do so once every ten minutes. It worked perfectly in the ICU but would later malfunction, resulting in me becoming severely addicted to morphine. How could the VA fail to notice that a malfunctioning machine had been injecting far too much morphine for more than a year? Morphine is a controlled substance. Are VA records kept so sloppily that no one takes notice of how much morphine is being dispensed? Although it could happen conceivably under combat conditions, such negligence should never be allowed to happen in a domestic hospital administered by an agency of the United States government.

At least I'm in good company. I read somewhere that Hermann Goering, the Nazi founder of the Gestapo, was also addicted to morphine. Do birds of a feather really flock together? Since Hermann Goering died three years before I was born, I guess we can only speculate.

I don't particularly like alcohol; I don't even drink coffee. I stayed away from drugs in my youth. Now, here I am, at 61, addicted to morphine. My long, curly hair hasn't been cut in five years and I'm stuck in a wheelchair. Once I was a Military Policeman. You might say I was a role model. Then the VA covered up by telling my son, a Major in the U.S. Army, that I addicted myself to morphine and that led to me losing his respect. I'm surprised that they could pull it off. You can't buy morphine on the street. Only the VA and the military have morphine in significant quantities.

Nothing could be further from the truth. But I have no recourse because veterans can't sue the VA. I sincerely hope that the people who read this book will strive to make changes in the system. I believe that the good citizens of the United States cheerfully pay for the VA, thinking that the money is well spent on caring for veterans. They deserve to know what's really going on: quackery, waste, and corruption.

Since morphine isn't very common, I probably should say a little more about it. Morphine is derived from giant poppies whose sap is collected and dried to make opium.

Processing opium farther produces morphine which with farther processing produces heroin. It's sort of like refining sugar:

Sugarcane - molasses - raw sugar - granular white sugar
Opium poppies - opium - morphine - heroin

What is it like to be on morphine? First, let's get something straight. Morphine pills mess up your stomach and won't get you a good high. In order to get high on morphine, it must either be injected or administered intravenously. Within minutes all pain vanishes and the patient feels comfortably numb. An increase in dosage may make the patient a bit queasy. With time, constipation ensues, but that can be corrected by drinking milk of magnesia or taking a laxative.

How many patients in ICU are on morphine? Automatic injection machines are pervasive. Morphine (at least at the VA) is fast becoming the "feel good" drug of choice. Keep them high on morphine and they won't cause trouble. Happy patients are not constantly ringing for a nurse and consequently require less care. So what if they are temporarily vegetables? If they survive, they can be taken off drugs before leaving the hospital. And, if they die, at least they die happy. What's wrong with that? What is wrong is me and guys like me who continue to crave morphine long after we get off of it. I don't just like morphine, I LOVE it. Thank God it isn't available on the street because I cannot remember a time I felt better than when I was on morphine. Administrators who addict veterans in order to save money should be fired. Instead, they are promoted.

For 66 consecutive Sundays, a group called the Veterans Revolution has protested outside the Los Angeles Medical Facility at Wilshire Boulevard. They are demanding that the VA administrators quit making policy about what is best for veterans and start asking veterans what they want. This is the United States of America. Because we fought for you, you are free to do as you choose. It is only fair for veterans to demand freedom, too. Get rid of the administrators and let the physicians and veterans determine policy. Dr. Feel Good can go take a flying leap. I want to walk. I want to drive. I want to live as well as everybody else. If an administrator needs someone to take morphine and be a gimp, he is welcome to have my wheelchair. There is nothing wrong with me that cannot be fixed. All I need is surgery minus the quackery.

Enough negativity. From what you have read or heard about the VA, you may have gotten the mistaken idea that the VA is rotten and corrupt. Though that may be the case for a few bad apples, the vast majority of VA employees (including healthcare professionals) go out of their way to assist disabled veterans. If they are off the clock and they see a veteran in a wheelchair struggling to open a door or not being able to reach something, they drop what they are doing and cheerfully come to his assistance. I have found this to be true of almost everyone both in the VA and the public at large. Taxpayers provide a generous pension to service-connected disabled veterans like myself along with free medical care and benefits.

I am extremely grateful to the public for all they have done for me. I went to school on the G.I. Bill. Later, I enrolled in the veterans' writing project which gave me an award for one of my short stories. Overall, the VA has been very helpful. My primary purpose in writing Quacks is to expose the inefficiency, waste, negligence, and redundancy which plague the VA, so as to improve performance and assist the accomplishment of mandated objectives. Constructive criticism is an essential part of the feedback process which government uses to determine how well it is doing its job. Quacks is intended to function like a report card; nothing less, nothing more. I apologize in advance for any connotations that are perceived to be destructive.

Etched into the entrance of the Department of Veterans Affairs (VA) in Washington D.C. is a phrase from Abraham Lincoln's Second Inaugural Speech: "To care for him who shall have borne the battle, and for his widow, and his orphan." President Lincoln obligated the nation to care for the men and women who have honorably served this country. More than 140 years later that promise still stands. My father and my mother served as President and Vice-President of the Fleet Reserve Association, Long Beach, for more than two decades. As a youth, I watched them visit disabled veterans in VA hospitals, comforting those who needed it. I was born at the VA hospital in Long Beach, California, in 1948 when it was a Navy facility. Of course, I had no idea then that I would come to depend on this and other VA hospitals for my healthcare. My parents were selfless role models of whom I'm justly proud. To the extent possible, I have followed in their footsteps in my support of veterans and enlisted military personnel. What my parents used to do on an individual basis, I now do online and in books so as to reach a far greater number of people.

Since ICU is akin to solitary confinement, it should never be used as interim warehousing of recovered patients awaiting further processing. The cost is enormous; my extended stay in ICU cost almost a thousand dollars a day. If the decision maker had to pay for it out of his own pocket, this wouldn't be happening. But he is paying for it out of your pocket and the pockets of everyone else who pays taxes. A few seconds of thought might have solved the dilemma. Administrators are paid to think. Those who cannot need to be transferred to menial positions commensurate with their skills and abilities.

The ICU physicians and nurses gave me drugs to suppress my immune system to keep my body from rejecting the artificial knee. Consequently, I was more susceptible to infections and similar disorders. The housekeeping staff did an outstanding job of combating germs, mopping the floor with disinfectant three or four times a day. But the sheets were changed on an infrequent basis and I have been told that they are laundered at a central facility that fails to ensure that the water temperature meets or exceeds 180 degrees for one rinse cycle so as to eliminate staphylococcus.

The Government Accounting Office should investigate and determine the rate of infection in VA hospitals and whether taxpayers are getting enough bang for their bucks. Is the rate of infection higher than in private for-profit hospitals? If it proves higher, the VA must be made to improve its standards and procedures. Care and treatment for veterans should not be permitted to slip to slipshod substandard quality. When the Inspector General finds out that something is amiss in the standard operating procedure of the Department of Defense and/or its hospitals, heads begin to roll. This should also be the case with the VA. Regular examination of VA hospital statistics needs to be undertaken by the GAO and compliance with standards strictly enforced.

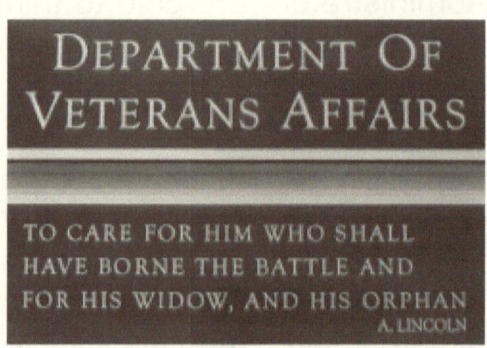

DEPARTMENT OF VETERANS AFFAIRS

TO CARE FOR HIM WHO SHALL HAVE BORNE THE BATTLE AND FOR HIS WIDOW, AND HIS ORPHAN
A. LINCOLN

Jerry L. Pettis Veterans Administration Memorial Medical Center, Loma Linda, California, featuring the duck pond, nursing home, emergency room, eye care facility, dental, and four floor hospital
11201 Benton Street, Loma Linda, California 92357
Toll free phone: 1-800-741-8387

Got a nasty habit? You can't get by for two weeks in ICU without a stiff drink and a pack of Camel cigarettes? Anything can be had for the right price. Usually, a patient or a nurse will offer to serve as a go between. Smuggling liquor and cigarettes into hospital wards is a time-honored VA tradition. Formerly, manufacturers gave away cigarettes to veterans in VA hospitals. Currently, smoking has gone out of style, but at one time field rations came with cigarettes and matches. If I remember the ads correctly "more doctors smoke Camels than any other brand of cigarette" because "there isn't a cough in a carload." The Surgeon General can put that in his pipe and smoke it.

Give the guys in ICU a big one finger salute and be glad that you aren't them. Who wants to be tethered to an IV and/or a catheter? Want to piss in a bag that hangs from the side of your wheelchair? Hell no! The best care is prevention. Don't wash down your Viagra with a shot of tequila, don't flavor your beans with bacon grease, and keep out of unlit brothels and dark alleys. Believe me, ICU in a VA hospital isn't for you. Your elected officials get to go to Water Reed. You get Loma Linda. They're smart and you aren't. But don't worry about it. If you fail to survive, the VA will bury you for free and mark the site with a solid brass plaque, commemorating your military service. That's right, you get to rot along with your military buddies. There is no finer way to go. A copy of your DD214 form gets you a deed to eternity.

I was born in a VA hospital and someday they will bury

me in a VA cemetery. How about that? I am getting cradle to grave coverage and I am a confirmed capitalist. Oh happy day!

If you are going to be admitted to ICU, take everything you will need with you. Take your laptop along and it is liable to get stolen. Besides, you won't be able to get on the internet with anything other than a Blackberry. The structural steel in the walls interferes with cell phones and the VA has yet to discover Broadband. The food is atrocious. Bring cash with which to bribe the kitchen staff into providing you with edible fare. Otherwise, a taste bud transplant should be scheduled with your surgery. Ever try toasted cardboard? *Bon appetit gastronome.*

When you find yourself feeling far too frisky for an ethical man in his golden years, when your spirit surges and overflows its banks, when life's cornucopia threatens to overwhelm you, a short stay in ICU might just be the reality check you need. Watching other veterans space out on mind-numbing drugs, watching people die before their time due to quackery, negligence and incompetence, and listening to the whir, beeps, and clicks of monitors that mostly monitor the best time to pull the plug, will certainly bring you back to your senses. Nothing is more illustrative of the human condition than pain and suffering.

Chapter 4

1 Southeast (1SE)

The vast majority of patients are discharged from the hospital when they finish ICU. But I chose to complete the six month physical therapy rehabilitation program so as to gain full use of my titanium prosthesis. Dr. Gustafson had told me that I was not likely to get more than 70 percent usage. However, the physical therapists thought otherwise. Their bonuses were based on performance. They said that if I worked hard enough, I could be restored to 100 percent. I wanted to be a whole man, not a pitiful gimp in a wheelchair. A ray of hope was all I needed.

There are three nursing homes on the first floor of the VA Loma Linda hospital. We don't talk about them. They are tucked away beyond the "employees only" kitchen and can best be reached via a passageway between the eye clinic building and the main hospital. Please refrain from calling 1 South, 1 Southeast, and 1 Southwest the funny farm or the cuckoo's nest because someday you might end up there, too, vegetating through marathon reruns of Jerry Springer and Maury Povich. Feel like killing your wife? Fancy a career as a rapist or a serial killer? Does Osama bin Laden ring your bell? If you answered yes to any of these questions, then you are a candidate for 1 Southwest where they test the latest tranquilizers and hallucinogens on veteran guinea pigs. My neighbor came back from Vietnam with all sorts of flashbacks. The

Emergency Room admitted him for observation and subsequently wheeled him past the double security doors into a fantasy world called 1 Southwest where they shot him full of happy juice and strapped him to a gurney. He emerged from there as a different man, totally devoid of will. His family now has to lead him around by the hand because the VA gave him the chemical equivalent of a frontal lobotomy.

This poor veteran also got a heavy dose of Agent Orange while fighting in Vietnam. Dow Chemical could care less. They should be held responsible for the effects of their dioxins. Hundreds of children in Vietnam and the United States have been born with birth defects. God rendered a rainforest which Dow Chemical tore asunder. There is no justification for defoliation. Why has the World Court at the Hague in Holland failed to adjudge this a war crime?

1SE is the entry port for the nursing home. It is primarily for rehabilitation and recovery, the average stay being four months. They admitted me for six months of physical therapy, but I caught a staphylococcal infection and wound up spending two years there. Infections have always run rampant in hospitals. If you don't want to get sick, don't go to a hospital. Put diseased people together with patients with suppressed immune systems and you have a prescription for disaster. It is not in the patients' best interest to conduct healthcare from a central facility. Far better to have doctors make house calls. Black bags are back in style among médecins compétents. Long term doctor/client relationships don't just happen, they have to be developed. Recluse physicians are a thing of the past. Today's doctors are built to last. They are part of the communities in which they live. Thus, they cut down on stress and anxiety. Responsible physicians do not accept gratuities from drug manufacturers and insurance companies. They sell their services, but not themselves. A true professional has character and integrity. By bringing back the Hippocratic Oath, doctors will regain respect by the public for the medical profession. One spots a quack by how it mucks around. Stop mucking and do the job right. God grants the physician the power of life and death, do not abuse it.

My respite in 1SE was organized and disciplined. At 7 AM, five days a week, the nurse at the front desk handed me a computer printout with my schedule on it. I was required to go to everything on the printout. Here's a typical schedule:

8 - 11 AM   Physical Therapy
11 - 12 Noon   Recreational Therapy
1 - 3 PM   Physical Therapy
3 - 5 PM   Audiology

Our cafeteria serves lunch at noon. Each patient had a tray that came from a compartment in one of the two stainless steel slop wagons that the orderlies wheeled in at mealtime. No matter what is served, it is lukewarm, including the salad and ice cream. Everything tastes the same; no salt, no pepper, no seasonings whatsoever. Cardboard is king. The food is so bland that the rats refuse to eat it. Here, try my strawberry rhubarb fruit cup. The VA buys it from Vietnam. I guess it's payback for waging war against them. The Marines splatter it on the wall when the nurses aren't watching. The spots and stains closely resemble the fly specks found in real military mess halls. Our two obese nutritionists claim that the food is good for us, but you won't see them eating it. Once a week, I pay one of them to bring me a cold bottle of Pepsi from the PX. I treasure it because it is my last link with civilization. If Margaret Mead was still alive, she would most likely do an anthropological study on it. But she is long dead. Nobody cares. What is needed is a sign at the entrance similar to the inscription Dante saw on the seven gates of hell: "Abandon hope, all ye who enter here!" As in Hades, many enter, but few return. Disease, infection, and negligence take their toll.

I can honestly say that the food that is served in VA nursing homes is worse than the food served in military mess halls. Patients who had money and weren't bedridden bought their meals at the food court on the second floor. It wasn't just that the food wasn't seasoned, it wasn't prepared properly. Many of the processed food items were offbeat and ended up in the trash can. Anybody for strawberry rhubarb fruit cup? Of course not. Why it constantly came up on the menu is beyond me.

Evidently, the nutritionists know what is best and are determined to give it to us whether we can stomach it or not.

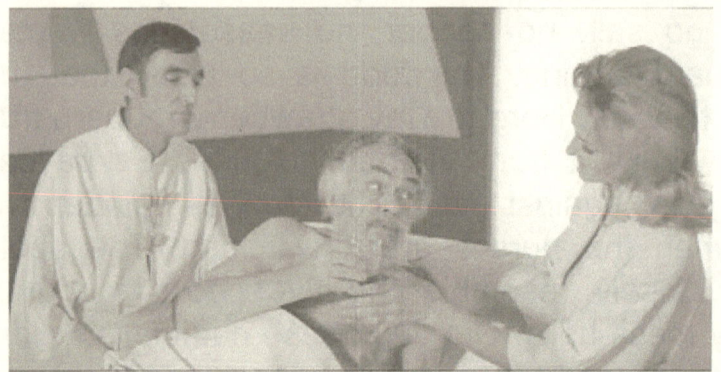

With such atrocious conditions prevailing, one must wonder why a veteran would want to enter a VA nursing home. Most have no other option. Some are dumped by their families. A select few view it as a way to hornswoggle an assisted suicide complete with burial and flag. If you are bound and determined to go, you might as well go for free like Sol Roth in the 1973 science fiction classic, Soylent Green, who was given a comfortable bed prior to having been prescribed a lethal dose of opiates and happy juice. Thank you, VA. People's Temple leader Jim Jones could have learned a thing or two from you.

And, Josie, the reservist Staff Sergeant head nurse of 1SE is uniquely qualified to teach it. Every morning she bumped against my sore right knee as she made her rounds. "Oh, did I hurt you? I'm so sorry," she feigned while savoring my excruciatingly painful involuntary reaction. Make no mistake, Josie is by no means a sadist. Rather, she believes it is her job as head nurse to maintain order and discipline in the unit. Although I made no overt effort to challenge her authority, she had heard me interpreting regulations for others, the gist of which she feared because she did not understand. Supervisors are promoted to the level of their incompetence. Torture is an ineffective tool whose use cannot be excused. No doubt the fires of hell burn in eternal vengeance for those who knowingly tormented their fellow men.

Having been designed in the early 1970's, the double occupancy rooms are quite spacious and have all of the amenities found in private facilities including a handicapped bathroom with a sliding door that is shared with the room next door (knock before entering). Rumor has it that it came into being as a result of the Sylmar earthquake in which 35 people were killed when two wings of a VA hospital collapsed. But please do not worry as there is not much chance of it happening again. Loma Linda VA hospital was built of steel reinforced concrete. However, if by some weird quirk of fate, it pancakes in an earthquake, the worst place to be is at the rear of the ground floor where the nursing home is located.

Most patients eventually succumb to infection. The infection begins as a small yellow dot that when left untreated expands into a pressure sore. If the sore is discovered and treated, it goes away. Otherwise, it becomes infected. Shifting bedridden patients helps to prevent bedsores. In fact, there are hospital beds available that automatically turn the patient as he or she sleeps. But the VA has yet to authorize the upgrade. Could it be that a burial plot in a national cemetery costs less than a decent bed? Sol Roth had it better. I suspect this also explains why unsuspecting wheelchair bound veterans get foam egg crates to sit on rather than thick gel cushions. What it doesn't explain is why physicians permit this to happen. I recall a young Air Force veteran, a well-respected and well liked long-term resident of the nursing home who over the course of several decades received a number of skin grafts for pressure sores on his rear. Following each procedure, he had to lie on his stomach until he healed. Obviously, it wasn't a joy for him, the VA, or the plastic surgeon. But this is what happens when medical schools teach by rote. Physicians who can't think are incompetent. Far too many doctors fall into this category. Next time you are in a medical school, watch an instructor and his students walk single file down a hallway with the teacher in the lead. This works fine for a mother duck, but a system based on privilege can't teach a physician to think. Those grueling 24 hour shifts that interns are forced to work do little to stimulate their brains. Zombies don't think and zombies are prone to make mistakes. Zombies don't belong in the medical profession.

The last hour of a nurse's shift is devoted to handwritten reports, none of which makes its way into patients' digital records. During this time, the ongoing shift is supposed to cover for them, but in my experience it rarely happens. I know of patients who have waited over an hour for assistance after pressing an emergency alarm. What's wrong is that the alarms are visual rather than audible and can consequently be ignored. Most nurses have come to regard the last hour of the shift as "their time." Once the reports are written, they sit in the nurses' lounge and gossip. How do I know? Like President Bill Clinton, they occasionally leave the door open.

In addition there are potlucks, sports pools, and other events, most of which take place on paid time and all of which serve to distract nursing personnel from taking care of patients. I got the distinct impression that the sale of handicrafts and jewelry provided a substantial source of income for many of the nurses. Patients were largely an inconvenience that they would rather do without. The solution is to give them drugs and tranquilizers so you don't have to deal with their problems. As far as much of the staff is concerned, the sooner they can pull the plug on you, the better.

Think I exaggerate? The turnover is so fast that names are written on masking tape before being placed on a room's door. Here today, gone tomorrow. How sad. Veterans don't rate a label maker. A former Command Sergeant Major (E-9) in 1 Southeast wrote a history of his service (numerous tours in Vietnam, etcetera) and attached it to his room's door. Most of us thought it added a humanistic aspect, but the administrators were afraid it was going to get out of hand. They eventually evicted him. The last time I heard, the Sergeant Major was living on the street in his truck. During a robbery, he was hit over the head, suffered a severe concussion, and the surgeon put a metal plate in his skull. It seems to have affected his balance as he now walks using a hand carved staff. Anyway, it made me wonder. How many combat tours does it take to gain the respect of VA administrators?

Doctors sometimes provide referrals to private specialists with the implication, spoken or unspoken, that the for profit specialist does a better job than the one that the VA provides. Are there kickbacks involved? This is something that needs to be looked into by the VA's Inspector General.

In VA healthcare veterans are assigned a general practice doctor who acts as a gatekeeper, deciding when a specialist is needed. The only way anyone can get an appointment with a specialist is through his primary physician. Patients in the nursing homes are not assigned a primary physician. The only way they can see a specialist is if the director of the nursing home refers them. In practice this rarely happens. In other words, nursing home inpatients living on the first floor of the VA hospital have a harder time seeing a specialist than they would if they were outpatients living at home. When, following an infection, my artificial knee was removed, I was constantly trying to obtain an appointment with the surgeon, but couldn't get one. In fact, I wasn't able to see him until after I went home. What good does it do to live in a VA hospital nursing home if you cannot be seen by a specialist when you need one?

I should mention that all of the rooms are exactly the same. Once you've seen one, you've seen them all. I confess that Stalinist construction does not appeal to me, nor do Daly City and federal housing projects. Just because something is egalitarian does not mean it has to be monotonous. In my opinion, long term residents should be encouraged to redecorate as long as it doesn't interfere with the staff's routine. For instance one of my roommates was a World War II Marine with a Purple Heart who had a large family. During visiting hours, our small room was packed with his relatives, most of whom had to stand because there is a rule against taking chairs from other rooms, even when they are vacant for a long period of time.

The three nursing homes on the ground floor of the VA hospital share a large rehab center between them that is stocked with exercise machines, weights, mats and other gymnasium items. It is staffed by upwards of ten full and part time rehabilitation technicians. Upstairs, on the second floor, there is an almost identical rehab center for outpatients. The 2nd floor rehab center also seems to be responsible for testing to make certain that the claims of the other center are not being fudged, which, together with the intensity of the rehab program, makes me think that rehabilitation technicians' pay is to a great degree based on how well their patients progress. In most cases this is a great idea, but it should not be done without the aid of a doctor monitoring the patient's overall condition. My rehabilitation technician worked hard on improving the arc of motion in my artificial right knee to the exclusion of everything else. Eventually, my calcified left ankle collapsed, erupting in a staph infection that spread like wildfire throughout the hospital and did not come under control until they sent me home to die. I didn't ask for the role of Typhoid Mary. It could (and should) have been prevented. Now, the surgeons won't fix me because they are scared to have me back in the hospital. I cannot walk or stand. Although I'm not to blame, they condemned me to spend the rest of my life in a wheelchair.

Rehabilitation was an arduous, albeit necessary, process in which atrophied muscles were toned and made to function. Rehab technicians chart your progress from the first faltering step to (ideally) full and complete recovery. My relapse was an anomaly. Rehabilitation is successful more often than not. I say this in all honesty; I am the exception that proves the rule.

Anyone who receives an artificial knee goes to rehabilitation. At private hospitals rehab is recommended but not required. At VA hospitals, however, there are no exceptions. Either one takes rehab on the second floor as an outpatient or on the ground floor as an inpatient. Because the taxpayers are paying approximately $50,000 for the operation, the VA makes certain that you do your part, i.e. rehabilitation, to make the outcome a success. It is either arranged as a package deal (artificial knee operation and rehabilitation) or it isn't done. At least in theory, there isn't any room for failure. I am living proof, however, that nothing is foolproof. Mistakes occur. Physicians are not immune from bad judgment. Surgeons must acknowledge and correct their errors. Attempted suppression only serves to exasperate the problem.

As previously stated, a patient's day is scheduled by computer printout. An individual can have up to four hours of rehabilitation in two hour accruements interspersed with arts and crafts, recreational therapy, aerobics, work therapy, and a litany of assorted subject matter that the staff have devised over the years, much of it of questionable relevance. Schedules come down from on high with absolutely no patient input. The attitude is "we know what is best for you." While that may be true for a few slackers who do not want to be bothered by the details of their treatment, the majority are offended by the VA's inflexible dictums. They appear to have been promulgated by an educator familiar with intermediate and high school scheduling in some bygone era when students automatically accepted the judgment of administrators in determining career choices. Why an adult veteran cannot be trusted to have a voice in his/her own treatment is beyond me. Big Brother has gone far enough. It's high time for change.

At the time of being admitted to a VA nursing home, the patient is asked to fill out a form showing whether or not he/she wants extraordinary means utilized to maintain life should a life threatening situation arise. What comes to mind is pulling the plug on someone who is brain dead and comatose. What doesn't come to mind is temporary dialysis, chemo and radiation treatments and an entire gamut of procedures that can be construed as artificially maintaining life. A roommate in 1 South became bloated and died because he had elected not to receive extraordinary treatment. Temporary dialysis might have taken the strain off his kidneys, but now we will never know. To my way of thinking, he deserved the chance he didn't get. God alone should determine when we die. Who wants the VA to play God?

If you would rather not have the government make life and death decisions for you, check the extraordinary measures block when you fill out the form. You can still change it later. It simply keeps the government from pulling the plug against your family's wishes. Don't ever authorize the VA to determine when the physicians should give up on you. There are enough conflicts of interest going on at the VA without you adding to them.

Remember the old roach motel commercials, "roaches check in but they don't check out." That is how it is with a VA nursing home. I term it the Hotel California syndrome. Private nursing homes are populated by patients who check in and check out at will. Once you are admitted to a VA nursing home, you must stay until they discharge you. Remember, when it comes to decisions regarding your health, Big Brother knows what is best. Born with an independent streak? Then VA healthcare might not be for you. You call the shots for only so long as you pay the bills. Once the VA becomes involved, you no longer run the show, so think carefully about it before calling on the government to help you out. And don't be surprised if they ask you to disclose your finances. You may have fought for freedom and capitalism, but now you are stuck with charity and socialism. It is best to stay with private healthcare if you can possibly afford it.

The chokepoint is at the front desk. Nurses determine at what point a specialist will be (or will not be summoned) for patients. Patients whose families ask about their condition and visit them regularly tend to receive better treatment. Moaning and groaning won't get you anything more than a tranquilizer and some pain medication. Feeling queasy? You are not the only one. Between midnight and 6 AM there is a semicircle of wheelchairs behind the front desk of patients who cannot sleep for one reason or another. Don't ask for a sleeping pill because you aren't going to get one.

Recreational therapy has a few things to offer. Wheelchair bowling, however, isn't one of them. Using ramps to play gutterball isn't much fun. Much better to accept a dinner offer from a fraternal order such as The Elks or the VFW. I have gone to some that were prime rib affairs with baked potato and decent beverages. It doesn't cost a cent. These guys will show you a good time. They genuinely care about disabled veterans. If nothing else, it's a break from the nursing home grind. Recreational therapy also has books on cassettes with walkman style recorders. It kept me from joining the wheelchair semicircle in back of the front desk at night. When I couldn't get to sleep, I listened to a John Grisham novel about evil corrupt lawyers and Wall Street Go-getters. Audio books kept me sane.

The packaged peanut butter and grape jelly sandwiches

with no crust are nourishing. With luck you can subsist on them for days. Avoid brands other than Smuckers. Put 20 or 30 in the meds refrigerator near the front desk and you can get a snack late at night. Microwave popcorn is another good idea. Heat it for 15 seconds longer than it says on the packet. Get kettle corn or buttery flavor. Stash it with your socks or underwear and never do it while a nurse is around. Nurses are good for smuggling alcohol and cigarettes and little else. They are all look and don't touch. Pontius Pilate didn't wash his hands half as much as they do.

Want to leave for a while? Good luck. Your family must sign you out on a pass. Ask your son to give the front desk a number at which you can be reached. Be punctual or the nurse won't ever give you another pass. It's just like the military, only better. Remember to take a shower before you go because you will have to pass inspection by the nurse in charge at the front desk. Sick and tired of rules? Isn't that just too bad? Without the VA, you would probably be pushing a shopping cart loaded with your belongings down the sidewalk. Try to be grateful.

I'm going to give you an idea of what your day will be like. At 4:30 AM a nurse shows up to administer my meds. She wakes me and my room mate from a deep sleep by turning on the bright overhead lights. She is positively bubbling over and couldn't wait until we have something in our stomachs to absorb it. Ducosate on an empty tummy will trigger an acid reaction. You are permitted to wonder how this ditzy drug pusher airhead made it through nursing school, but please have the courtesy not to ask. When it is 6:30 AM, we shuffle to the cafeteria for breakfast. Oh boy, it's strawberry rhubarb fruit cup for the fourth day in a row. And a great big helping of mush. Bet that will act like a catalyst on the Ducosate and the other meds the nurse made me take at 4:30 in the morning. Time to go to the restroom and get ready for the 7:30 AM lineup at the front desk. I even have time to go outside and feed the ducks the hard roll they gave me at breakfast. Last week the Director complained that our bread crumbs were making the pond turbid. That's too bad because I don't have a quarter to put in the gumball machine that dispenses duck food. Administrators who don't like it can eat cake. Down with the Bastille and up with the ducks in the moat. Revolution is brewing albeit remote.

A revolution is the farthest thing from the mind of a per diem medical professional. Try to imagine these people in Cuba making $45 per month. They are dedicated to accumulating a sizable bank account and little else. Sin vergüenza. My ex-wife became a money hungry registered nurse. Only a money grubbing nurse would abandon her son. May she choke on someone else's sausage.

Ever had a turkey hot dog? Dry with a plastic skin. Almost inedible. The nutritionists like them, but you won't. No matter how they are cooked, they taste like cardboard. Yum yum in your tummy. These things must be close to 90 percent fiber. The turkeys may have died of old age, but you won't. Too much of this kind of food will most likely give you a colon or gastro-intestinal disorder. No more granola for you. Wouldn't you rather have a shot of 200 proof rum and a crack whore? It's only a matter of money. When you get rated at 100 percent, it's like a cardinal's robe in that it will bring out the best veal when you go out to dine in Los Angeles and/or Las Vegas.

Do you need a bed bath? Only student nurses give bed baths. The riper you get, the better the chance of disease and infection. Hello, I thought this was a hospital. I've seen cleaner kennels. These people need a bath or a shower. Please, give them one. What's wrong with the night shift? The handicapped showers sit empty at night. Get on the ball. Where is the DAV and the VFW when you need them? This place is nasty. They claimed I got infected with staph by not washing my hands. My hands aren't the only things dirty. Please, get rid of the filth and corruption.

Once a baby is born, you cannot reinsert it into the mother's womb. The same maybe true of the elderly and disabled. Once you enter a nursing home, you will never reenter society in the capacity in which you left. You have become a marked man. Sooner or later disease is going to take you down. Pills and alcohol cannot solve the problem. Get a new lease on life by developing a raison de vivre.

If you are off balance and/or unsteady on your feet, the nurse is supposed to put a wireless perimeter alarm on your bed at night, but it never seems to get done. The nurse could care less if you fall because that would mean one less patient. Stan, my roommate died at night and fell on my missing knee, causing me to scream and wake up the other patients. That's too bad. When you complain about waking up to witness a compatriot's passing, you will probably pass without a whisper too because you have foretold your own death by being a magnificently insensitive jerk.

They used to have colonies of lepers, where people dare not venture. Am I diagnosed with a leprous disease? Or is it simply that nobody wants to come and visit me? It is Christmas and everyone else is having a feast. Here I am in a VA nursing home dining on coagulated macaroni and cheese. When I was in the Army, I did much better than this. The cook tried hard to cook us a hot ham Christmas dinner. The Holidays were full of cheer, why can't it happen here? Scrooge must be in charge of Veterans Affairs and scrimping on the food as if it was being paid out of his own pocket. I barely weigh 116 pounds. I am emaciated and my ribs show through my skin. Is it too much for me to ask for a hot Christmas dinner?

Wheelchairs these days are made of plastic, one size fits all. But I had to lean over to grab the rim and brakes of a 20 inch wide wheelchair. Needless to say, I almost never got to go anywhere. In fact, it barely fit through the doorway in the bathroom. Worse yet, I couldn't get into the meditation room of the chapel. Must I die for my sins before my time? Let me linger a little longer. I may yet surprise everyone by doing something stupendous with my life.

Ever had a catheter? You are in for a treat. A masochist will ram a tube up your penis. Your job is to scream. No Vaseline for you because you are a macho Marine. Look on the bright side; your male nurse is gay and he enjoyed every minute of it. Isn't that a swell plastic bag? Please don't spill it. Have yourself a Big Gulp and then you can fill it.

Do you like shoes? Are you aware that some kinky people have a fetish for them? I guess this explains why the VA has us wear green wool socks with black tread on the bottom. Socks without shoes makes a fashion statement. The VA asks you to wear the socks once and then throw them away. Don't wash socks, Beau Brummell would never stoop to washing socks. Taxpayers have money to burn.

Aren't you cute? You are wearing a flimsy gown. The ties are in the back to keep you from undoing them. Now you can squat to urinate. Isn't that wonderful? The VA gave you a gender change without you asking for it.

Lost? Can't find your way? That's because all of the corridors in the hospital look alike. Formerly, veterans followed colored tape to get to various locations. That system has given way to one of naming the corridors. Ornately painted street signs on the walls of 1 Southeast denote the location. It isn't GPS, but it's the best the VA could do.

Need a copy of your medical records? All VA records are stored in digital format. You are entitled to a CD copy but must request it in writing. Even if you don't want it, it is a good idea to get one just in case you are asked for it by a private doctor sometime in the future. It's free. All it requires is time.

Men and women veterans are housed separately. There can be no conjugal interaction. The sick and the elderly don't have sex.

Every evening there are DVD movies screened in the mess hall and the microwave popcorn is free. This is done by the patients and the movies are not censored by the staff. Overall, I enjoyed them.

The physical therapists do their work well. However, the doctors fail to monitor the patient's overall condition. Infection often sets in without them noticing it. I know of at least three veterans who contracted staph infections without anyone noticing it. Spending time with nursing home patients should be a priority for doctors, but they seldom come downstairs to examine us.

In two years I was examined less than five times. VA physicians are negligent and incompetent. I am in a wheelchair because of them.

The first rule of the Hippocratic Oath is "to do no harm." But the VA surgeons cannot be bothered by such trifles. They are much better than the rest of us. It must be nice to be able to get away with anything. Rules are for stupid people, not for prima donnas who have special skills.

Supposedly, a veteran has a right to have an issue investigated by the Inspector General. However, the Inspector General does not investigate the vast majority of complaints. Are you so naïve that you expect to obtain justice from an in-house investigation? You would probably do better to take the case to court or inform the media.

Likewise, patient representatives usually prove to be a waste of time. Nobody cares about veterans in nursing homes except for the veterans themselves.

Few deceptions are as cruel as those of in-house investigators. They are as liable to find fault with the agency that gave birth to them as I am to criticize my own dear parents. It does not make common sense, yet that is exactly the way it is.

I never saw nor heard of a patient stealing from another patient. However, my black leather jacket was stolen by a VA employee. VA hospitals have their own uniformed police departments. Still, I don't think it is a good idea to bring anything valuable with you when you are admitted to a VA nursing home.

Most of the corridors have hardwood railings. If your wheelchair is too wide, they can greatly speed you on your way. Therapists recommend egg crate cushions, but if you can afford one, a gel cushion is less likely to give you a pressure sore.

The VA also has a drug recovery program for veterans that they recruit from the street. These "Silver Spoons" help feed patients who cannot feed themselves. They also do gardening and other forms of physical labor.

Football pools are circulated among the patients by the medical staff. The person starting a 10 x 10 pool may take three or more squares as his/her "rightful share" for organizing and circulating the pool. You should be aware that these pools are not officially sanctioned by the VA and sometimes turn out to be scams.

My ex-wife is a RN at a private hospital, but she used to work at Riverside County hospital on Magnolia in Riverside. I remember watching them load wheelchairs on a truck when they moved to Cactus Street in Moreno Valley six years ago. I never wanted to be in a wheelchair then and I definitely don't want one now. The VA claims I belong in a wheelchair, but I know better. It is only a matter of time until I walk again. Keeping a positive attitude is a big part of recovery.

The nursing homes have an arts and crafts workshop located in a room at the rear of the mess hall. Though therapists refer to it as recreational therapy, it consists of making moccasins without hard soles and leather wallets without compartments by running rawhide strips through holes on the edges. If you enjoyed doing this kind of thing as a kid at camp, you might want to sign up for it here. Fortunately, it's optional.

When someone leaves the nursing home, it is common practice for a junior nurse to cram their belongings in a clear plastic bag. Ask to have it double bagged as I have seen the bag punctured on the way out. Picking your stuff off the cement isn't fun, so be forewarned and benefit from other patients' mistakes.

When someone is authorized to pick them up, patients normally wait for them at the front desk. I have seen patients wait for five hours. I suggest you go ahead with what you normally do. They will tell you when your guest arrives and you won't be disrupting your schedule any more than necessary.

Near the front desk is a big refrigerator that is meant to be used for medicines. However, a number of patients keep snacks in it. Because the VA hospital Food Court closes early, hiding food is almost a necessity. There aren't many things worse than having your stomach growl at night and not being able to do something about it.

During the holiday season, private religious elementary schools handcraft greeting cards for veterans in the nursing homes. The ladies auxiliary from a veterans organization crocheted blankets for us. I got a red, white, and blue one. Everyone was given two phone cards. Several days before Christmas, the DAV gave me an electric alarm clock. A number of singing groups sang carols at night in the mess hall. The plastic Christmas tree looked a bit ragged, but we made do with it anyway.

There is one part-time barber for the nursing homes. She has a retail shop on the second floor of the hospital. In two years as a patient I only saw her come downstairs once. We got shaggy in between visits. I no longer bother cutting my hair. Many barbers will not cut the hair of disabled veterans who don't have enough mobility to transfer into a barber's chair. They claim it's because of their insurance policies.

The amount of money you receive from the VA is determined by your percentage. Percentage is a function of how bad you were injured in military service and whether it was service connected.

Though the VA offered to get me an electric wheelchair, I chose to stay with a manual because the muscles tend to atrophy and you tend to put on weight in an electric wheelchair. Getting (and staying) in shape is an important part of survival. I use ramps to maintain my upper body strength.

Criticizing for the sake of criticism may be cathartic, but it isn't a substitute for detailed analysis. Personally, I believe that the VA specifically and government in general are slowly strangling the private sector by knuckling under to pressure groups and failing to enforce discipline. Promulgating regulations does little unless those regulations contain a penalty for non-compliance. It is my observation that situational ethics are utilized to justify mistakes while the private sector must compete in order to survive. Thus, private business is forced to change with the times while the VA too often perpetuates errors in order to protect its image. Image is not nearly as important as transparency. There is no need for whistle-blowers when institutions function efficiently. Why cover up for quackery, negligence, and malfeasance? Doesn't the VA have better things to do? The budget needs to be keyed to how well the VA does its job. Only then will there be true control over the VA.

Current patient representatives who necessarily depend on the VA for their salaries and advancement need to be replaced with real ombudsmen who are independent of the VA. I would prefer these to be supplied by veterans organizations and hired by the GAO to keep tabs on the VA's performance. Giving the VA more money without making them accountable is a waste of taxpayer resources.

The VA hospital has a Credit Union which both veterans and VA employees can use. Although few inpatients make use of it, it is a way of receiving monthly electronic deposits without having to get a check in the mail. It made managing bills easier while sick in a hospital. I chose to award my son my power of attorney so that he could conduct my monetary affairs for me. I must say he did an excellent job of it.

Microwave popcorn is a staple in the VA nursing homes. It does not cost much and it is easy to make. Kettle corn is my favorite. Kettle corn is both sweet and salty. There are also varieties with butter and other flavorings.

You might as well resign yourself to it, fully ninety percent of the inmates will die while living in this place. Although escape is not impossible, it is highly unlikely that you will have the balls to pull it off. Remember, when you entered this institution you gave the administrators your power of attorney. They own you. How does it feel to be a slave? Your plastic wristband brands you a part of this nursing home. All that awaits you is a VA cemetery plot and a brass marker.

As you acquire more prescriptions, you increase the chances of the prescriptions not being compatible. The pharmacist will help you with this. Also, some prescription drugs deplete the body of certain substances such as folic acid. If you aren't sure, you will have to check with the pharmacist in order to find out.

Those flimsy slippers they issued you will not get you anywhere outside of the hospital. They will most likely fall apart at the first puddle. Doom stalks you. You were sentenced to life for nothing in particular. Nonentities suffer a fate worse than death. You are wearing a gown and slippers. It's Tinker Bell in drag. Stay in the hospital because the rest of the world is laughing at you. A joke, but not a very funny one. Feeling depressed? You have no inkling of how much you depress others.

You look like an experiment that went wrong. When did you last change your underwear? Take a shower and scrub your armpits with lye soap. Put on a fresh gown and join the party. Party until you puke. Scumbags like you don't know when it's time to call it quits.

You can vote in elections while you are in a nursing home, but it is your responsibility to register prior to the election. Even if you cannot get out of bed, you have the right to vote. I voted in both the senatorial and the presidential election. Most of the patients voted. It is important to remember that we are a democracy and we have the right to change the government by voting the bums out.

Remember, you must share the bathroom with others. I found it better to shave and brush my teeth before going to bed. Time in the bathroom is limited in the morning. Also, we had a man with an intestinal problem in the next room who sat on the toilet seat for a long time. There are also times when patients require help with transferring.

It is best to memorize your blood type. My doctors gave me two transfusions while I was in the nursing home. Your blood type is on your dog tags and also on your medical records. My blood is A positive.

Because I could not transfer very well, they gave me a

bed bath using two basins of warm water and a bar of soap. I washed my genitals myself. Student nurses were usually designated to take care of the patients while the staff nurses disappeared and took care of personal business. Nurses should be rated on the basis of how well they take care of patients rather than how much the administrators like them. Many of them don't speak English well and don't get along with the patients. If you complain, they label you a racist and nothing gets done about it.

It takes money to survive in the nursing home. If you don't bribe the nurses and orderlies, they will make your life hell. Interest is charged at the rate of 100 percent per day. Do your best to stay out of the way. Once you have been there you will never call it a rest home again.

Can anyone tell me what the surgeons did with my knee? Since the VA surgeons removed my knee, I haven't been able to walk. If the VA refuses to give me another artificial knee, they can put back my original since I was able to walk straight-legged with it. The first rule of the Hippocratic Oath is "to do no harm." The VA has a responsibility to restore me to my original condition. What can be done to get the surgeons to correct their mistakes? This is the result of treating surgeons like prima donnas. We have to answer for our errors. Why should surgeons be any different?

Because I can't stand up on the scales, I haven't been weighed by the VA in more than six years. Although the

VA has a special chair that weighs patients that can't stand, they hardly ever take the time to use it on me nor have I ever had water treatment for my pain. Warm whirlpool baths would definitely help.

Collectivized healthcare is a mistake. The VA would work better if it issued vouchers to veterans that made them responsible for their own healthcare. I would like to have healthcare as good or better than other Americans.

There was one PC for use by patients in the game room across from the cafeteria. A great time for utilizing the computer was at night. After several months of trying, I discovered how to upload files to the domain, fdungan.com, I bought prior to knee surgery when I was writing my first novel. With this I communicated with a number of people who were sympathetic to my plight. Without a website, I would still be lingering in 1 Southeast. Please, don't underestimate the power of the internet to level the playing field to where anyone can expose government incompetence, waste and corruption. Truth is omnipotent. Injustices can be corrected provided you work diligently on them.

When Congress passed the Americans with Disabilities Act, the rationale was for individuals in wheelchairs to be able to access buildings, both public and private. However, most buildings built prior to the law have yet to be renovated. This is very frustrating to people like myself.

Since the advent of the all-volunteer military, veterans no longer matter to the government. We are Uncle Sam's cast-offs. Every once in a while they throw us a bone out of guilt. We are elderly and soon we will be gone.

On August 19, 2004, I was discharged from 1 SE by Doctor Bull and was subsequently driven to my home by my son, Che.

Chapter 5

Home I

After a number of months went by, the rehabilitation technicians arranged to rate my artificial right knee's utility at 102 percent, a remarkable figure, considering the 82 percent reading it had the week before.

Regardless of how the figure was obtained, I was released from the nursing home the next day without taking any blood or urine tests. My personal items were stuffed in a clear plastic bag by a nurse and they telephoned my son to pick me up.

But I was in pain. My physical therapist had told me to put most of my weight on my left leg. However, my left ankle was hurting. After being home for two weeks, I found my left ankle oozing an orange fluid. The surgeon on duty at the VA's Emergency Room immediately used a scalpel to slice both sides of my ankle open to drain without giving me a shot to numb the pain. I yelled like I had never yelled before. It was loud enough to hear in the lobby and my son rushed in to see his father in agony. The date was September 18, 2004.

Upon examining the fluid, they found that I had staphylococcus, which had entered my bloodstream and had caused my body to reject my artificial knee. In an operation that night, the surgeons removed my titanium knee and substituted a spacer that left my right leg in a permanently bent position. It would affect my sleep and I could no longer stand or walk.

Later, I was sent to ICU where I was given two transfusions, put on an intravenous drip with antibiotics, and attached to a device that gave me a little morphine every 10 minutes when I pressed a button.

I was in ICU for 9 days when they transferred me to the nursing home at 1 South on the first floor of the VA hospital. At the time, I was not aware that I would be there for over a year.

While I was at home, an older woman charged me $50 to clean my house for approximately two hours. I also ran through seven different gardeners in two months. Many of the gardeners knew even less about gardening than I did.

Chapter 6

1 South

Subsequently, I was transferred to the nursing home on the first floor, 1 South, where I was given a strikingly similar intravenous drip with antibiotics and morphine. A nurse inadvertently put the wrong settings on the morphine device and I received two times as much morphine than I was supposed to get. This went on for a long time and nobody noticed. Unconscious more than half of the time, I missed many meals and lost nearly thirty pounds. By and by, they found out their mistake and tried to blame it on me.

The VA programmed its own television whereby the patients got to vote on most of what was shown. They opted for a mixture of Jerry Springer and Maury Povich shows. Three or four hours of back-to-back Jerry Springer Shows were followed by almost the same amount of Maury Povich Shows. I found it hard to believe that Jerry Springer had formerly been an alderman in Chicago.

Many of the patients in 1 South seemed to like watching people throw chairs at each other. Perhaps it's because they didn't see much action in the nursing home. The Jerry Springer Show was choreographed in the same sense that professional wrestling is not real. The Maury Povich Show featured welfare mothers and the men that fathered their babies. Five or six of the men would be DNA tested and often none of them came up positive.

When this happened, derisive shouts would occur. I must confess that I, too, was upset by the outcome. Most mothers can identify the man who fathered their child. Surely, Maury Povich displays the extremes at the expense of the norm. Racially biased programs are not permitted on television. Unfortunately, the same doesn't appear to hold true for shows that are derogatory to a particular gender. Evidently, exploitation remains the rule despite protests to the contrary.

1 South is one of the VA's best run nursing home. Many private healthcare facilities do little more than warehouse their patients and string them out on tranquilizers. It's a disgrace how invalids are treated in the United States. In Asia they have more respect for their elders than we have in our culture. At least, that is what I have been told.

The 1 South and 1 Southeast nursing homes on the first floor of the hospital share the same game room, kitchen, cafeteria, and rehabilitation center. There the similarities end. They each have their own administrators and the staff is different. Since 1 South is a tad more isolated, it has been designated to warehouse the long-term patients. I am lucky in that I survived to tell my story. I trust that you will learn from it and not make the same mistakes that I did.

1 South had an excellent administrator with an open door policy that kept things running smoothly. This man really cared. On his only day off he brought his wife and son to sing karaoke with us in the cafeteria at night. Instead of

the usual popcorn, we had a barbecue that he paid for with his own money. It was not lost on the patients that this man led by example. Although I have been discharged for a number of years, I can still recall how much he cared. His exemplary behavior and judgment was a rarity at the VA. Why can't the other people in authority be like him? Is it too much to ask for people at the VA to do a good job? It's common knowledge that people regularly get away with sleeping on duty at the VA.

It is best to get along with your roommate. There is only one TV per room and which show to watch is not always agreed upon. I often went into the game room or the cafeteria when I started to feel frustrated because of my roommate's inflexibility. Life is too short to be getting continually upset. If your blood pressure gets too high, it can hurt you.

1 South is connected to the new eye clinic building, but patients rarely travel from one building to the other because it is open to the elements in between. Perhaps the VA is planning to enclose the area. If they did so, it would make 1 South far less isolated.

While I was in 1 South, an anonymous donor gave some books on tape to recreational therapy plus walkman-style recorders on which we could listen to them. I went through a number of John Grisham novels that had been placed on cassettes. By the time I was discharged from the hospital I had gained an appreciation for modern novels.

Veterans tend to eat at tables with other veterans who served in the same branch of the military as they did no matter when they were discharged. This holds true in the retail food court upstairs as well as in the nursing home's mess hall. All tables look pretty much alike and there are no signs marking them, but everybody tends to sit in the same place every time. You soon learn where to sit because if you sit in the wrong place, the others won't talk to you. This is especially true of Marines. Despite a lot of empty chairs, nobody but Marines eat at the Marine table.

Most people are familiar with non-profit organizations that serve as advocates for disabled veterans, such as Disabled American Veterans (DAV) and Veterans of Foreign Wars (VFW). They are either free or have small yearly fees and have offices which can help you with the paperwork and documents you will need. A lot of smaller organizations such as the Fleet Reserve also exist to represent you, however, only the big ones have representatives at the Loma Linda hospital. You can also get phone cards along with shawls, pens, and writing paper from them. The Red Cross dispenses donuts, bagels, and coffee to patients without charge (they take donations). The DAV contributes a van to the VA and a local auto dealership donates the hospital electric shuttle cars that pick up people from the parking lots. Volunteers operate an escort service that pushes manual wheelchairs and book carts.

Despite 1 South having an exemplary administrator, it

wasn't all that different from 1 Southeast. Due to regulations promulgated in Washington D.C., any one person in the bureaucracy doesn't have very much latitude regardless of his/her position. Probably this explains why change comes so slow.

The VA's nursing homes are run more for the benefit of the staff than for the patients. At least it seems that way. There are more cases of patients being treated as objects than I can relate. The elderly have feelings, too. We aren't guinea pigs and we should not be used as test subjects without our permission such as the Tuskegee airmen were used in studies done by physicians.

Vouchers would be one way of putting veterans in control of the VA. That way veterans could choose their own services, making them competitive. However, if the vouchers were ever devalued by the government, they might not be sufficient to cover costs. I doubt that vouchers would provide a permanent solution. I don't think dismantling the VA would be a good idea, but the veterans need to be in control and there must be some way of doing this.

The VA also has therapy dogs provided by volunteers that carry their ID cards on a thin chain around their necks and have been fixed. They entertain and interact with patients. It brings a touch of the outside world to the nursing home. These dogs are docile and are a great hit with the patients.

The VA nursing homes have too much paperwork. Forms aren't as important as patients. The VA hierarchy keeps creating more forms as time goes on. This has got to stop. Burdening the staff with paperwork has gotten out of hand and patient healthcare is suffering from it.

It is rare for two doctors to agree on anything. They aren't really different from the rest of us, other than they are better educated than most of us. Always get a second opinion on an irreversible diagnosis. It cannot hurt and most physicians encourage you to do so. I certainly wish I had taken the time and trouble to obtain a second opinion on my right knee before undergoing surgery. I was not aware that the surgeon was inflating his success rate.

The medical staff is scheduled with three shifts of eight hours, a day shift, an afternoon shift, and a graveyard shift. After 10 PM, half of the lights in the hallway are turned off while the late night shift is working and most of the patients are sleeping. There are bombproof heavy metal doors on the outside entrances that are locked at night. Also, the new Dell computers which the medical staff uses record the keystrokes so that administrators can view emails and other documents that have been written.

I once tried to speak to the hospital's Director in his office about the warehousing of nursing home patients and was informed by the secretary that there was no permanent Director and that the position remained unfilled for upwards of three months. I ended up making the complaint to the secretary. How can anything get better when there is nobody in charge?

Toward the end of my stay the staff sold us kibbled duck food in an attempt to keep the patients from feeding the ducks the hard rolls the patients received at dinnertime. Because the rolls were almost inedible for old patients with few teeth, we soon returned to feeding the rolls to the ducks, along with the kibble. As far as I know, nothing has changed and the kitchen staff is still serving hard rolls with the dinners. But a parking structure will be taking the place of a large portion of the duck pond. The VA is going to tear down paradise and pave it with a parking lot. It reminds me that the VA had valet parking at one time and it failed miserably.

The State of California DMV has license plates and placards for the cars, trucks, and vans of disabled veterans, but the hospital has inadvertently obstructed the process by requiring additional paperwork. Disabled person plates are much easier to get.

If you are going to be admitted to the nursing home, bring some underwear because the gowns are flimsy and it is easy to catch cold because the temperature is kept low by the nurses in order to prevent disease. My thermostat was stuck and I almost froze with no underwear. It takes the repairman at least a week to act on a work order.

Very sick veterans belong in a nursing home, but the rest would be better off at home with medical surveillance.

Nursing homes, due to their crowded facilities and vast numbers of patients with reduced immunity, tend to be hothouses for diseases. I became stronger when I went home. Why not care for elderly patients at home? It costs far less to assist them in their own home than to send them to a nursing home. If the VA implemented it, it would undoubtedly make budgetary sense.

Although doctors don't know it, the Dell computers that they are supplied with are part of a central network which makes a dated record of each and every keystroke. That way if something is to be questioned an administrator can examine the record. Little is gained by it. There is something about getting a printed readout which an administrator finds it hard to resist. I believe that there is too much stealth going on at the VA. This is not the American way which countless veterans have given their lives to defend.

If you are in a wheelchair, like I am, you may be having difficulty reaching things. Most dressing sticks have a small cup hook on one end and a large clothes hook on the other that have proven helpful to me. However, don't be embarrassed to ask a stranger to get an item for you from the top shelf at the market. A pipe or a long pole can also prove useful. Devices with suction tips can also help extend your reach and maneuverability.

As you accumulate prescriptions, there is a chance that one pill will conflict with another. If in doubt, ask for an opinion from one of the pharmacists. It is part of their job to inform the physicians of drugs that work at cross purposes. You can also renew drugs online, by mail, or by calling a toll-free telephone number. If you can possibly do so, please renew them 20 days prior to running out of a particular prescription. Reorder early as there may be a backlog during certain periods such as the holiday season.

The VA has gone electronic. Physicians now make entries via a Dell network computer. Nothing is handwritten. Congress wants healthcare providers, both public and private, to go digital.

1South has a top loading washing machine and a dryer in a tiny utility room for those patients who prefer to launder their clothes themselves. The VA has a laundry which cleans towels, sheets, gowns, and washcloths without charge to patients. These items are provided by the government and are accessible to patients.

If you are stuck in a wheelchair in a nursing home, you won't be able to do pushups and sit-ups, but you can add muscle to your upper body by using a manual wheelchair on a daily basis to go up sidewalks and ramps. Establish a routine and do not deviate from it. Eventually, it will become second nature. By transferring you can also build strength.

Some of the pills may be too large to swallow. If this happens to you, ask a nurse for a pill-cutter. However, both halves will have jagged edges which can scratch your throat. The physician said to have a nurse pulverize the pill and then mix it with my food at dinner. I have a difficult time keeping my dinner down. I'm afraid that I will vomit on the nurse.

If you take a wrong right turn in 1 South, you will go through the double doors into 1 Southwest, the psycho evaluation center for the VA hospital. It is also a hospice. Patients there are strapped to gurneys and administered tranquilizers. This is where the VA puts veterans with psychological disorders. My neighbor across the street spent several years there due to a disorder relating to the Vietnam War. The sounds that come from there are pathetic and moving. Security would not allow me to enter and I have no intention of returning.

Pressure sores are the biggest cause of infections. Sores begin as a tiny yellow spot that doesn't hurt. Use a water-based gel to get rid of them. If you are a male in a wheelchair, buy a cushion to keep them off your genitals. Apply talcum powder liberally. An eggcrate cushion can only be used when it is dry. I suggest you obtain a thick vinyl-covered cushion. It's important because you will be in your wheelchair for a large part of each day.

Painkillers can affect your regularity. Codeine and morphine will result in impaction when you take them for a long time. To avoid an enema, you will have to cut down on them. Milk of magnesia and similar laxatives are merely a temporary solution.

A ramp van featuring 4-point attachment can be provided by the VA at your request. The vehicle comes complete with driver and is free of charge. It is made available to veterans in wheelchairs who are unable to transfer to vehicle seats.

The handrails throughout the ground floor are made of polished hardwood and mounted low to accommodate wheelchairs.

The VA also has a string of satellite nursing homes for veterans in Barstow and a number of other cities. However, I preferred to be at Loma Linda due to its rehabilitation programs. In addition, the State of California and many other states operate homes for veterans. Private nursing homes and assisted living is available for those who can afford it. Like the rest of U.S. society, the gap between care for the rich and for the poor is enormous. Relying on the government to take care of you in latter life is definitely a mistake.

In 1 South women are segregated from the men. 1 South is like a time machine that got stuck in the 1950's. It's hard to imagine a more conservative institution than the VA. Above the entrance to the hospital, there are four large American flags. A painting of the current president dominates the lobby. A framed copy of the Declaration of Independence and other patriotic documents line the opposite wall. An old style outgoing mailbox with a schedule is around the corner. The only thing modern is an ATM machine that stands to the left inside the front entrance.

The chapel offices are at the back of the building on the ground floor. There are Protestant and Roman Catholic clergymen. The Roman Catholic mass is at 9 AM on Sunday and the Protestant services are at 10 AM. Everyone is welcome to attend.

There are a number of tiled roll-in showers in 1 South. Cover all open sores with plastic bags waterproofed with adhesive tape. I usually shower at night when the line isn't as long. Transfer to a PVC chair with wheels before taking your shower. Benches and shower chairs simplify the process. Bring your own soap. Wash cloths and towels are provided by the VA's laundry service.

The pull-up metal triangle hanging on a chain above your pillow makes an excellent antennae. Reception inside the VA hospital is terrible. Stand near the window in the daytime if you desire to listen to a battery powered FM radio without interference.

All of the windows in 1 South are one-way in which patients can see out, but the public cannot see in. This is better than Candid Camera and reality television combined. Spy on doctors without them knowing you are watching. Be aware that there is a risk in using binoculars. Keep records with times and dates. Blackmail administrators and get better fed. Obtain lots of information and you can become powerful. Play your cards right and internment in 1 South can become an opportunity. Always make the best of what you have. Be active and you will get better. Leave 1 South behind and go on to lead a full and productive life.

The inpatient pharmacy is smaller than the outpatient pharmacy and it is at the other end of the hallway on the ground floor. The hours are also different. Why the inpatient pharmacy is shielded by clear bulletproof plastic is a well kept secret.

When you express your frustration with the VA's one-size-fits-all manual wheelchair, your social worker will probably offer you an electric wheelchair. Electric wheelchairs chew up doorways and entrances. They lead to obesity, flabbiness, and feebleness. Try to get a better quality titanium wheelchair that is lightweight and serves as an extension to your body. Their gel tires need no air.

The rehabilitation therapists also issued me a walker which I no longer need because I cannot walk. In my office it is simply one more dust catcher. Why not help the budget by having veterans give back items which they no longer need? Then they could be reissued to veterans who need them. It simply makes too much sense for administrators to implement it. Bureaucracy demands that all items are purchased new. Consequently, they waste our tax dollars.

When a patient becomes a threat to himself or others the nurse is supposed to set up an electronic perimeter alarm that notifies medical personnel when the patient leaves his bed. This device often malfunctions and is not often used. Nevertheless, a nurse is held responsible for not using it, so they often lie or cover up. They are almost never fired. Unless a complaint is filed, nobody cares if patients die like flies. The death rate for 1 South can be attributed to a lack of concern among most nurses and doctors. Elderly patients simply don't matter.

Some floors are better suited to wheelchairs than others. Those of 1 South are covered with large squares of linoleum which are far better than other materials to make manual wheelchairs roll. Nor do wheelchairs leave unsightly marks on it. The hallways of 1 South are where I learned how to use a manual wheelchair.

The biggest sin in 1 South is to scream at night and awake your roommate and the other patients who are sleeping. Rather than finding out what caused you to scream, the staff always blames the person who screamed. Patients suffer in silence because of the strict discipline. The staff gets away with it because they are unionized. When the entire nation gets federal healthcare, don't say that I didn't warn you that it is wasteful and abusive.

The VA has an outreach program for homeless veterans. These veterans are given a bus ticket to the nearest VA hospital where they are enrolled in a drug rehabilitation program. They help out in the nursing homes as Silver Spoons. From what I have seen, they improve in health and have a low rate of recidivism.

The beds in 1 South are extremely narrow and the rooms which are occupied by two people are not as big as most of the rooms which accommodate a single person in private hospitals. Wives and other relatives are treated dismally. Those who want to stay with the patient overnight must rest in uncomfortable armchairs. Children's hospitals have superior facilities for relatives. The VA is insensitive and obtuse.

When patients can't take a shower, a student nurse is assigned to give them a bed bath. I received a bed bath once a month. In between baths I got rather stinky and was extremely unsanitary.

Clear vinyl wristbands are worn by nursing home patients which contain, among other information, their social security numbers, leaving patients open to identity theft. Clearly, the VA should put a stop to this hazardous practice.

Since the gowns are flimsy, it is best to wear decent underwear. Boxer shorts result in less chaffing around the groin than jockey shorts. Talcum powder also helps to prevent chaffing. I prefer to use 100 percent cotton t-shirts, either V-neck or crew. You don't need more than six sets of underwear nor do you have room for more.

Toothpaste and other toiletries, with the exception of razors and blades, are provided by Disabled American Veterans (DAV) and other veterans' non-profit organizations. The cart usually comes around several times a month to each room with paper, pencils, pens, and other useful items which are provided free of charge. I belong to the DAV and I can tell you from experience that they make good use of their resources.

The VA is a bureaucracy. Many of its employees do not function well. Either others take up the slack or the job doesn't get done. A substantial number of bottlenecks exist. It is frustrating to see how inefficient and ridiculous the VA can be.

Disposable urinals and bedpans are made available for patients who are too weak to make it to the toilet. Also, these containers come in handy when the bathroom is occupied for a long period of time.

The worst part about being in a VA nursing home is that most of the staff treats the patients as if they were charity cases. I could have fled to Canada during the Vietnam War, but I didn't. Those who did owe veterans a debt of gratitude. Our former president, Jimmy Carter, was wrong in granting amnesty to those who fled to other countries to avoid the draft. And he was wrong in giving back the Canal Zone to Panama. People who get something for nothing rarely appreciate it.

The water container on the nightstand next to the bed is usually refilled twice at night with ice water. If you require more, you will have to replenish it from the tap by yourself. Please do it quietly and avoid waking your roommate.

Several times I saw snowy egrets and other large water birds in the duck pond. Later, I learned that the duck pond was a part of the Pacific Flyway for migratory birds. Currently, the VA plans to remove part of the duck pond for a parking structure. Why can't the parking structure be constructed in a location where it won't have a devastating effect on the environment?

There is an upright piano at the far side of the mess hall. I once heard a professional musician play it, but that was the only time in two years that I heard it.

1 South issues script which is recognized by the Retail Store on the second floor. The script is worth one dollar and can be used to buy anything the PX has in stock. Because the PX is federal, there are no state taxes. Frequently script is awarded as a prize for winning at bingo and similar recreational activities.

Towards the end of my stay in 1 South, I went on one of the few outings I was allowed to go on during my two years at VA Loma Linda. Recreational therapy drove a bus to a local bowling alley and we attempted to knock down pins by rolling the bowling ball alongside our wheelchairs. Also, ramps were available for those who needed them.

Loading the bus for the trip to the bowling alley was harder than the recreational therapists anticipated. A chrome security "bar" rose from the hospital's loaner wheelchairs and the wheelchairs did not fit into the bus. The therapists tried and failed to remove the chrome anti-theft devices. In the end, patients who used the hospital's loaner wheelchairs were unable to take the bus to the bowling alley.

Boxer shorts often ride up on males in a wheelchair. It is best to purchase knit boxer shorts with a button on the middle of the fly that keeps the fly from inadvertently opening. They are sold in a variety of styles, sizes, and colors.

VA therapists give patients with limited range of motion a variety of devices to help them cope with their disabilities. Semicircular knives that rock assist patients who have difficulty grasping with their hands to cut and slice meat. Patients like myself who can't reach their feet use nylon straps to pull on their socks. Dressing sticks enable patients to retrieve objects and dress themselves.

Due to my limited range of motion, I usually dress myself in knit boxers and a T-shirt. I have lots of pants and shirts, but I cannot spare the time that it takes me to get ready. I would rather go to church in a suit and a tie, but it is better to go without a suit and tie than to fail to attend services.

There are drinking fountains with stainless steel basins at many points along the hallways of 1 South. They dispense cold water through spouts at two preset heights.

I went to a number of aerobics classes. For disabled patients in wheelchairs, there are leg lifts and arm lifts which heighten your range of motion. Muscles can atrophy if you do not use them.

The citizenry meant for the VA to serve the interests of veterans and their dependants. Unfortunately, it currently serves vendors and employees. They skim off the top and suck the life from the VA. The bureaucracy of the VA is full of corruption and greed.

Throughout the year, the nursing homes are visited regularly by private Christian elementary schools. Prior to Veterans Day and Christmas, the students make cards for each of the patients. As far as I know, the public elementary schools have outings which do not include Veterans nursing homes. The patients are lonely. For the most part, the patients feel isolated from the community and enjoy receiving visits from children. After doing their best to serve their country, they wind up in a nursing home. Thank God for the children. I looked forward to their visits. The future of our country depends upon them.

Chapter 7

Home II

1 South eventually removed my IV drip, I began to go to chapel, and the VA hospital discharged me. Because my right knee had been removed and I could not walk or stand, a VA social worker rated my disability at 100 percent, service-connected, and gave me the services of a wound nurse who visited me regularly until the incision where the VA surgeon removed my right knee grew new skin over it.

I have now been home over five years. Although I have become stronger and I qualify for surgery, orthopedics refuses to correct what is wrong with me and the VA will not insert a titanium knee in my right leg. I cannot stand or walk and can only transfer to a bed or a toilet. The VA has purchased me a 2007 large van with an automatic side door and ramp for which I am grateful. I have added a second battery to the van to make the electronics work properly.

Earlier, I had bought a 2006 Ford Taurus with power brakes and power steering from a local dealer. Its V-6 engine ran well and it had only 37,000 miles on it. Also, it got good gas mileage.

Due to my open incisions, I received a visit from a wound nurse periodically. She removed the staples. I received the services of a caregiver from the VA for twenty hours each week. They were my arms and legs. I would rather be doing things for myself, but I must face reality.

Because I was unable to use a hose effectively while sitting in a wheelchair, I hired a contractor to put in bubblers and sprinklers that watered all of my plants and trees. It has two timers, one is located in the front and the other in back. Although I can't stand or walk, I maintain an orchard and a garden. Fresh fruit is much better than the fruit sold in markets, plus I do not use pesticides.

As an outpatient, I was surprised when Dr. Barton (who was the decision maker for the VA Loma Linda hospital) announced that I would not be receiving a new artificial knee and I would stay in a wheelchair for life without being able to stand or walk. The VA would not be operating on me despite the fact that I walked into the hospital and their surgery made me worse.

Because of the pain from where the VA removed my right knee, I am unable to sleep more than two hours at night. The doctor's response has been to prescribe me more pain pills. I don't want pain pills; I want a new artificial knee. As far as I am concerned, I have spent more than enough time in a wheelchair and the VA ought to fix me so that I can walk again.

I bought a non-aggressive black female AKC Labrador to serve as my assistance dog. She helps me by retrieving items for me, as well as by pulling my manual wheelchair through the soft dirt in the empty lot next to my house. I regard her as a good friend. She is extremely loyal.

I listen to the radio, read non-fiction books, read the newspaper and stay active to keep my mind off the pain. If I took half of the drugs the VA gives me, I would be addicted to them. The way to get rid of the pain is to replace my artificial knee. The VA cannot feel my pain, therefore they don't care about it.

Because I cannot stand or walk, I have to eat vegetables, fruits, and salads to keep from becoming impacted. When I was in the hospital, the doctors prescribed me milk of magnesia in order to have a bowel movement. Now, I keep myself regular with fiber.

I only sleep for several hours at night. Most of the time, I cannot sleep due to pain in my right leg. Painkillers no longer help. The VA should do the right thing and provide me with a new artificial right knee. Otherwise, I risk becoming addicted to pills, in which case I will be of no further use to myself or others. I desperately want to get well.

My desire does not seem to be shared by the VA. Although they have performed the same surgery on other veterans, they won't do it on me. I have written the VA Inspector General concerning Loma Linda's reluctance, but cannot get the decision reversed.

I had a man remove the carpets from the hardwood floors in my house, which made my wheelchair roll better.

Linoleum was put in the kitchen, hallway, and utility room and the bathrooms were tiled. I cannot go in much of the backyard because it is soft dirt.

I have to vote by mail because I cannot go to the polls. I have a permanent exemption. I permit politicians whom I support to put signs on my lawn as a way of declaring my loyalty. For 33 years I have been living in this house and the 5th Ward in Riverside is very important to me. I have seen this neighborhood change for the better in the three decades I have lived here.

I bought the black neoprene bandage with Velcro fasteners that I wear on my left ankle on the internet because the VA does not have it. Nor do they carry my wheelchair's gel cushion nor other items that are required by disabled veterans. One way to fix this problem is to require purchasing agents to become familiar with the requirements of disabled veterans in order to get promoted.

Because I run the risk of becoming infected when I put anything over where the surgeons operated on my right knee, I have five pairs of shorts which I wear year-round and there isn't a blanket on my bed. Also, I cannot wear shoes anymore due to incisions on my left ankle. Since I cannot wear much clothing, it costs me a lot to heat my house. However, the Southern California Gas Company gives me a twenty percent medical discount which makes gas more affordable.

I had a general contractor build a 3 bedroom guest house in the backyard. The children's bedrooms are upstairs. To date, I have not rented it out. The housing market has gone bust throughout the world.

It is Memorial Day. At the VA, I see wounded returning from Iraq and Afghanistan without any arms or legs. They are undergoing rehabilitation and many are receiving prosthetics. I pray that the government will take good care of them. Some will be stuck in a wheelchair for the rest of their lives. I wish I could do something for them.

Due to my disability, I can no longer drive the 2006 Ford Taurus and have no other option than to drive the 2007 Chevy van that the VA generously purchased to replace it. Although I could use a brake button, the VA refused to pay for it. I make do with what I have. With two batteries, it is very reliable. I do not drive much because my right knee has been removed and I no longer drive as well as I did before the operation.

Because I do not need both vehicles, I am attempting to sell the 2006 Ford Taurus. However, to date I have not gotten any good offers. Perhaps, it is because I refuse to take payments.

I continue to work on the house. A new floor was installed in my bedroom recently. It is composed of tongue and groove 5/8 inch hardwood boards. Also, the kitchen faucet is being replaced.

I am putting a larger double insulated window in the utility room. Also, I am having the cracks caused by earthquakes repaired. I have hired a contractor to do it for me because I cannot walk.

I go to church on Sundays by driving the van the VA purchased. For more than a year, I have been a member of Allen Chapel at 10th and Locust in Riverside. They have helped me to cope with the pain without drugs. I am very grateful.

I cannot help but scuff up the doorways with my wheelchair as I go through them. Some of the doors have been removed. Since I live alone, there is no door on my bathroom. Now, I am able to get to the bathroom when I need to make a bowel movement. It might not enhance my social standing, but I am not going to get impacted in order to impress others.

The doorways have for the most part been enlarged. Much of the house has been repainted. Aluminum angle strips were cut for the corners of the drywall to keep my wheelchair from damaging them.

I have an appointment with Ortho on June 15, 2010. The VA will transport me to Loma Linda VA hospital by a hired hack and will give me my first chance at getting repaired since coming home. Arthritis has damaged several fingers on my left hand and I can no longer play the piano or type on my computer like I once did. My missing knee is frozen in place and is extremely painful.

I do not sleep well. Currently, I have a cold which I cannot seem to shake. Since I already take too many drugs, I am reluctant to take more. Being sick in a wheelchair isn't much fun. I wonder if the quacks get perverse pleasure in watching me suffer?

To find out I met with Dr. Gustafson on June 15, 2010, six years after going home. What a poor pathetic wretch. Why people like him become a surgeon is beyond me. He had difficulty scrolling a computer screen and seemed to be unmotivated. I'm sick and he won't fix me. Although he cut on my knee, he won't complete the job. If it weren't for him, I wouldn't be in this wheelchair.

The budget for the VA is scheduled to increase by 11 percent in the next five years. Nevertheless, the VA intends to leave me in my current condition without replacing my artificial knee. I might as well have gone to Canada when the government reclassified me as 1A in 1968. The VA has failed to keep its promises.

The VA surgical team does not correct their mistakes. They see no reason to do so and it is to their advantage not to operate on me. If I had been told the truth up front, they would not have cut on me. As far as I know, they are the only surgeons in the world who are not required to finish what they started.

Several years ago, I asked my primary care physician to please send me to Walter Reed Army Hospital for evaluation. I couldn't even get that out of her. My new primary physician is better, but not by much. What do I have to do to get the VA to fix what they did to me? Why not give me a new titanium artificial knee? Am I ever going to receive what is due me?

I am scheduled to see the dentist at VA Loma Linda tomorrow. I sincerely hope my crowns turn out right. I am looking forward to being able to eat a good meal.

I'm now able to eat meat without hurting because the VA dentist installed gold crowns on my molars. As a way of celebrating the occasion, I have ordered a set of steak knives from a vendor on the internet.

I spent the morning adjusting sprinklers and repairing damaged stucco. These things are relatively easy because they are close to the ground. I can't climb a ladder and must hire an electrician to replace the light bulbs that are beyond my reach. This is very ridiculous. If the surgeon would replace my artificial right knee, I could change my light bulbs myself. Before undergoing surgery, I was very independent. Being in a wheelchair has taught me to be humble.

I was born at the VA hospital in Long Beach in 1948 when it still belonged to the United States Navy. Since I seem to have fallen between the cracks at the Loma Linda facility, I might as well try the Orthopedics department at the Long Beach hospital. I doubt that Long Beach would be as uncaring as Loma Linda. Anyway, I feel that it is worth a try. After all, they could not possibly be as callous and insensitive as VA Loma Linda has been.

My request to get a second opinion at Long Beach VAMC never was acted upon. Evidently, like so many other veterans, they let me fall between the cracks. I am righteously indignant. Records do not simply disappear. Obtaining a second opinion should not be so difficult. Quackery, incompetence, and negligence cannot be permitted to triumph. Congress intended for Veterans Affairs to serve veterans, but it has come to be the other way around.

Under the Obama Administration, the Department of Veterans Affairs has revived an old pamphlet for veterans that seemingly promotes euthanasia. When the pamphlet is read as a whole, veterans are left in bewilderment as to the government's valuation of their lives. Encouraging the practice of euthanasia from the federal government level, and specifically directing that encouragement at veterans, is disturbing.

On July 2, 2009, the Department of Veterans Affairs (V.A.) re-issued its revised handbook, Advance Care Planning and Management Directives. The handbook suggests the use of a pamphlet called, Your Life, Your Choices as a tool for drafting advance directives or living wills. However, Your Life, Your Choices when read as a whole, presents particularly disturbing encouragement to veterans to cut their lives short when facing even the mildest of difficulties, such as an inability to "shake the blues." Critics of the pamphlet have dubbed it the "V.A. Death Book." According to Jim Towey, former Director of Faith-Based Initiatives under the Bush administration, "it makes people feel like they're a burden and that they should do the decent thing and die." The Bush administration previously suspended use of the publication after researching its impact on "complex health and moral issues." However, under the Obama administration, the V.A. has again chosen to promote and encourage the use of the pamphlet.

Your Life, Your Choices encourages euthanasia. For instance, the pamphlet provides exercises in which one must ask whether his "life [is] worth living." Within this exercise, one must rate certain criteria as either "difficult but acceptable", "worth living but just barely", or "not worth living". The list of eighteen criteria to be rated outrageously includes the inability to "get around in wheelchair", the inability "to get outside", and an inability to "shake the blues". This list also includes other guilt-inducing criteria, such as an inability to "contribute to [one's] family's well being" or imposing "a severe financial burden on [one's] family." As Towey explained, the government begins sliding down a dangerous slope by assigning levels of value to life, for example, by implying that life loses its value if dementia sets in. Moreover, the V.A.'s promotion of Your Life, Your Choices raises concerns because its author, Doctor Robert Pearlman, advocates assisted suicide. In fact, the original version of the pamphlet referred readers to the American Euthanasia Society and the 2007 version of the pamphlet referred readers to Compassionate Choices (also known as the Hemlock Society).

While the V.A. has stated that they are "revising" the pamphlet, Towey considers this a mere attempt at damage control and criticizes the V.A. for not removing the pamphlet from their website as they are revising it. On August 23, 2009, Senator Arlen Specter, a member of the Veterans Affairs Committee, said that the pamphlet raises questions and is calling for an immediate hearing. Senator Specter also thinks the pamphlet should be suspended until hearings can be conducted to investigate the matter.

While the idea of congressional hearings is certainly welcome, the use of this pamphlet should not continue. The V.A., under the direction of the Obama Administration, should immediately cease the use of this pamphlet. The American Center for Law and Justice calls on the President to immediately intervene and permanently rescind the use of this pamphlet.

Because I have been in a wheelchair for an extended period of time, it offends me that they are implying that living in a wheelchair may constitute a reason to end one's life. Suicide is repugnant and should always be discouraged. A responsible government agency would not advocate such a policy. This constitutes proof that the Department of Veterans Affairs needs to be thoroughly revamped.

A federal bill that would require the Veterans Affairs

Department to post medical quality assurance records online is being fought by the V.A., which says it has concerns about the confidentiality of patients and whether disclosing screw ups online would have a "chilling effect" on the willingness of the hospital staff to report mistakes.

The legislation being disputed is HR 3843, the Transparency for America's Heroes Act, sponsored by Congressman Joe Sestak, a Democrat from Pennsylvania. If it gets signed into law, the bill would result in the V.A. posting quality assurance information to the internet after patients' identifying information is removed.

On September 29, 2010, Representative Sestak told the House Veterans Affairs Committee's health panel that the bill was necessitated by "revelations of substandard care" in the past two years that include well-publicized problems with the sterilization of medical equipment, and less publicized issues, such as a veteran whose open wound was filled with maggots and a diabetic veteran who was not given needed insulin shots while hospitalized.

Major veterans groups, including Veterans of Foreign Wars, Paralyzed Veterans of America and Vietnam Veterans of America, support the bill.

"Recent reports of contaminated instruments, unsupervised medical procedures and adverse conditions at a Philadelphia long-term care facility erode faith in the

VA health care system," said Michael O'Rourke of the VFW. "We believe that having information easily available to patients and stakeholders renews the emphasis on quality, accountability and sound health care procedures provided by all staff in every VA facility."

Carl Blake of Paralyzed Veterans of America said requiring VA to publish redacted medical quality assurance records on its website "will provide users of the VA a better understanding of the successes or failures of the VA in the quality of care they provide to our veterans."

"This may encourage greater efforts on the part of VA employees, staff and leaders to ensure the best care is provided while ensuring openness," Blake said.

The VA sees things differently. Dr. Robert Jesse, principal deputy undersecretary for health, said redacting names and other identifying information does not guarantee confidentiality — and the whole idea may hurt rather than help.

"We understand that some of the interest in transparency is to promote accountability," he said. "VA strongly believes that our employees must be held to the highest standard when delivering care. However, it is also imperative that employees know that they can report information fully and completely so that changes can be made and care can be improved."

Given the disagreement, the bill's fate is unclear. The health care subcommittee will not decide whether to move ahead with the measure until November 2010, at the soonest.

Epilogue

I did not mean to convey the impression that VA medical care is hopelessly flawed. I am, however, of the opinion that it needs to work on its priorities. Automated flushing toilets, motion-sensing paper towel dispensers, and similar showpiece "inspection tour" amenities should take a backseat to the VA's primary mission of providing veterans with quality healthcare. Administrators would do well to focus on improving standards rather than engaging in promotional gimmicks. Anyone who hires unlicensed physicians should be summarily fired and barred from federal employment.

A number of employees work for the VA because they could not make a go of it in the private sector. Since they are civil service, they are agonizingly difficult to discipline and/or terminate. Most earn considerably more money and receive better benefits than their counterparts in private industry. Personnel does a poor job of screening out the bad apples. Consequently, Veterans Affairs is inefficient, wasteful, bureaucratic, and far too much uncaring. Individuals become social security numbers. Patients are asked to fill out stacks of forms by clerks behind counters who stare at the floor as they dehumanize you by commanding "what is your last four?," never once looking up from their keyboard or calling you by your name. I am tempted to pull them out of their chairs, drag them across the counter, and make them acknowledge my individuality by looking into my eyes. At the bottom of my Social Security card it's clearly printed "not for identification purposes." I strongly agree. Can't these officious officials read? They make my blood boil.

Workloads for nurses at VA nursing homes are too big. Patients frequently must push an emergency alarm for an hour or longer before a nurse responds. This is ridiculous. LVN's complain that they perform the work and RN's get the money. Over two years, I watched as RN's sat in the Nurses' Lounge and the rest of the nurses took up the slack. This quite often happens at night after the physicians and administrator have gone home. All too often, when serious mistakes, including death, occur due to excessive workloads, the head nurse covers up for the other nurses in her report.

In 2007 in a Philadelphia VA nursing home a mute and disabled Vietnam veteran, David Allen, 56, choked to death on solid food the nurses fed him, even though they were directed to feed him a soft-food diet. Allen's sister, Belinda, said she was told the VA "did everything they could."

The Loma Linda VA nursing home employs two nutritionists, but the food is prepared by apparently unskilled cooks who seem to take little or no pride in their jobs. Inpatients had difficulty eating the meals; many chose to buy their food in the retail Food Court on the second floor or survived on food brought to them by their relatives from home. Upon being admitted to the nursing home, I weighed 141 pounds and by the time I left I had shrunk to 116 pounds because of an inferior diet. I was emaciated to the point where my ribs showed through my skin.

My immune system was suppressed. Bowel movements were a dark shade of green because the antibacterial drugs I had been given had eliminated the good bacteria from my body as well as the bad. My muscles had atrophied making it difficult to transfer from a wheelchair to the toilet or the bed. I suspected that I had been sent home to die and this was how the VA regularly kept a patient's demise from affecting their statistics. Appearances are more important to administrators than veterans' well being. Do I come across as being overly cynical? My surgeon has given up on repairing the damage he did to me. He has sentenced me to a lifetime of confinement within a wheelchair. That makes him a negligent, incompetent quack. I have no option other than to be wary of the federal government agency which employs him.

The Veterans Administration supposedly is run for the benefit of veterans and their families. Isn't this contrary to human nature? It would function better if it was run by veterans and their families. Some of the VA's employees are recruited in foreign nations who often have a negative view of the American military. They simply don't belong taking care of veterans and their families. Consequently, they tend towards marginal performance. Wouldn't we be better off without them?

Why do VA nursing homes have to look like an institution? After all, they are home to long term residents.

Why isn't suppressing a patient's individuality considered abuse? Aesthetics should be the purview of the veterans as long as they do not interfere with how the institution functions. What's wrong with hanging photos of loved ones on the wall? It is depressing to have all rooms the same color. Rooms too closely resemble barracks. Why weren't they designed with closets? Isn't this a quality of life issue? The VA should be doing everything it can to discourage elder abuse. Rooms should have broadband internet service. We need to be part of society. Our culture will be made richer by our presence.

Doesn't it make good sense to clean all of the fecal matter from an instrument that has been used to examine another veteran's prostate?

On April 3, 2006 the GAO reported that the VA evidently did not think so, and had not done it, because the manufacturer did not specifically tell them that they should. It appears that thousands of veterans' prostates were examined without having the probes that were inserted, you know where, cleaned before being used on other patients. The GAO discovered that numerous veterans had been exposed to potential infections including HIV.

Almost everyone employed by the VA is a member of the Public Employee's Union. Their counterparts in private industry are not paid nearly as well. In addition, VA employees receive generous benefits and automatic cost-of-living adjustments. It makes little sense that they have more activities and events scheduled than the patients in the nursing home. It would be more descriptive if the name of the department was changed from Veterans Affairs to Public Employee's Affairs. Although public employees have a right to make a decent living, they are not entitled to deplete the budget. I also maintain that the VA has too many administrators and not enough workers.

Security is lax. Identification cards are not being checked at the main entrance to VAMC Loma Linda. Inside, there is an ATM on the left. While it does not surprise me that the bureaucrats have put veterans at risk from terrorist intruders, it is odd that they do not seem to care about putting a freestanding ATM at risk. Train depots receive better protection. I would certainly feel safer with a security guard at the door.

A documented report released in December 2008 by the United States Justice Department maintains that residents of William F. Green State Veterans Home in Alabama suffered harm from the facility's inadequate medical and nursing services.

The report found that the nursing home had terribly "inadequate nutritional and hydration services." Further cited were abnormal psychotropic medication practices. Objectionable pressure-sore treatment and skin care; issues concerning restorative care and specialized rehabilitation services; failure to protect people from harm due to falls; failure to investigate all allegations of resident abuse; and inappropriate use of restraints were also highlighted by the report.

In November 2008, the VA Medical Center in Augusta, Georgia sent a letter to more than 1,200 patients who were treated for ear, nose and throat disorders, warning them they may have been exposed to infections. Considering that I spent two years in VA nursing homes due to an infection I acquired in a VA hospital, I find this particularly disturbing. Following surgery, patients are especially vulnerable to infection because their immune systems have been suppressed. Although VA administrators are well aware of this problem, they continue to pack these patients like sardines into Intensive Care Units and Nursing Homes where infections are extremely hard to control once they take root. The obvious solution is to quarantine patients with highly communicable infections such as staph and mercer, but in two years I never saw anyone, including myself, quarantined because of an infection. This amounts to abject quackery because the VA knows how to solve the hospital infection epidemic, but continues to say there is little they can do about it. I came close to dying and many of my fellow veterans have died from communicable infections, but the quacks at the VA Medical Centers won't do anything about it. They go along to get along. It is the height of moral turpitude. The very first principle of the Hippocratic Oath is to "do no harm." I call them quacks because they are quacks. I would sooner put my faith in the quacks in the pond than to put my faith in the quacks who work inside the VA medical center.

In 2008, the widows of two men who died following surgery at a Veteran's Hospital in Illinois confronted the hospital in court.

The two men died as a result of alleged substandard care while under the care of a surgeon practicing at the hospital. One man bled to death following gallbladder surgery, while the other died from a blood infection after the same surgeon did a lymph node biopsy on him.

The grieving widowed women claim that Illinois Veterans Affairs failed to conduct mandated background checks on the surgeon, who had been accused in Massachusetts of negligent practices in the treatment of seven cases in 2004 and 2005, two of which resulted in patient deaths.

Following the investigation, it was found that a minimum of nine deaths were directly attributable to substandard healthcare.

No more quackery. Unlicensed doctors are murdering veterans. They need to be terminated and prosecuted to the fullest extent possible under the law. I came close to being a victim. They are the enemy. It's our mission to rid society of these parasites.

Veterans Affairs is a "top-down" organization. The people at the top (who are not necessarily veterans) make decisions affecting the people at the bottom. Monarchies and dictatorships prefer it because it allows them to rule with an iron hand. However, such bodies are top heavy, inefficient, and do not respond well to our changing needs. Quacks, sycophants, and martinets abound in this type of bureaucratic agency. Exposing them is often difficult and terminating them is even tougher. Corruption is inherent.

The VA needs to be overhauled and decentralized. Participation by local veterans should be encouraged. We can do better than the professionals who spend more time enhancing their careers than fulfilling veterans' needs. Trimming the fat will enable us to accomplish more with funds allocated by Congress. No veteran should be left behind. Transparency is paramount. When errors are made, they must be acknowledged and corrected. No more cover-ups.

In the October 25, 2010 edition of National Review Online, critic Michael Tanner of the Cato Institute writes that the VA regulates costs by limiting how much it spends on medical care in a given year. "When resources can't meet demand…the VA rations."

Psychiatric and pharmaceutical services are limited by budgets, not based on patient need. The system is highly bureaucratized and when problems arise, "no one takes responsibility for fixing them," Tanner asserts.

I agree. Although it has been years since I was discharged from Loma Linda VAMC, I am still in a wheelchair because nobody is willing to take responsibility for fixing my knee. I am just another veteran. Why should they care?

Fixing problems involves more than making a name change. VA used to be an acronym for Veterans Administration. Then it was changed to Veterans Affairs. No matter what they choose to call it, the agency remains the same. It doesn't give us better health care or help us to live longer. Much better if the VA quit trying to pull the wool over our eyes. I would much prefer they make real improvements.

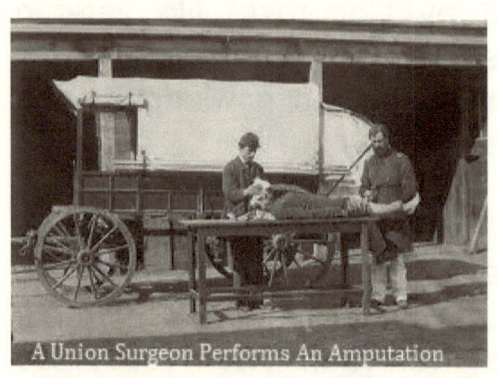
A Union Surgeon Performs An Amputation

During my two year stay at a VA nursing home, I was frequently treated as if I was a recipient of charity. Too often this was used to justify negligence and abuse. I had more difficulty scheduling an appointment with a clinic than outpatients do despite the fact that the nursing home is located within the hospital. The VA lost my dental bridge, but they would not permit me to see a dentist. Gumming my food was demeaning. It resulted in severe weight loss and weakened my condition.

At the end of the 19th century, a parcel of land in what has since become the extremely wealthy district of Brentwood near UCLA was donated to the federal government for building a retirement home for veterans. Over many decades, the property increased in value and is currently estimated to be worth $3 billion. Rather than build a nursing home, the Veterans Administration decided to lease the land to a private school for tennis courts, swimming pools, and a running track located behind a million dollar fence.

Because there are approximately 20,000 homeless veterans on the streets of Los Angeles, it is imperative that the land be used for its originally intended purpose. Currently, homeless veterans sleep on the sidewalk outside the locked fence.

A protest supported by veterans organizations has been staged outside the locked gate every Sunday for two years. Greed and corruption by the VA and government officials are keeping them from that which is rightfully theirs.

Vietnam veteran Wiley S. Drake, 64, drove up from Buena Park in Orange County to join in the protest.

"There is nothing wrong with beautifying Brentwood," said the Baptist preacher. "But not with veterans' money."

Opposing the protesting veterans is an organization which calls itself the Veterans Conservancy whose membership consists of wealthy Brentwood homeowners. They own television networks and movie studios and make large scale donations to influential politicians. Despite numerous financial scandals and cover-ups, they manage to avoid incarceration. It seems they don't have to follow the law because they can afford to buy their way out of it. The demonstrations have gone on for two years now. It is about time for the veterans to succeed. Money has the power to bring about change, but it cannot transform a wrong into a right.

Veterans healthcare suffers from unbelievable bloat. Resources are not keeping pace with increasing enrollment. Consequently, waiting periods for non-emergency appointments have grown to the point where a veteran sometimes has to wait months to see a specialist. Currently, a veteran must schedule an appointment to have his teeth cleaned by a dental technician fully six months in advance. That's ridiculous. Automatic paper towel dispensers equipped with motion detectors are nice, but the funds could be better spent on improving healthcare and dental services. What the VA desperately needs is to establish realistic priorities for its budget.

Mission creep is a consequence of having an overly

centralized bureaucracy. Washington D.C. regularly issues new regulations entailing additional responsibilities and duties for local Veterans Affairs medical centers without giving them the necessary funds to get the job done properly. When an undue strain is placed on healthcare resources, cracks begin to appear in the system and patients fall through them. Nobody notices because everyone is concentrating on complying with the new regulations in an effort to avoid being disciplined. My case has been put on hold by the orthopedic surgery team because fixing my knee will do nothing to improve their statistics. In the healthcare game, the trick is to go with the new and forget about the old. Images can be ruined by acknowledging one's mistakes. The VA shouts its successes from the rooftop while clandestinely burying its victims. I choose to wait because I am in no hurry to get a bronze marker with my name on it.

Several years ago, when Dr. Gustafson's colleague, Dr. Barton, informed me of the VA's decision not to rectify their mistakes by giving me a new artificial right knee and doing something about my fused left ankle, I requested to be examined by a orthopedic surgeon at VAMC Long Beach in order to get a second opinion. My request was summarily denied. Such is the arrogance of VA orthopedic quackery.

Unlike private surgeons, VA surgeons don't have to worry about being sued when an operation goes wrong. Rather than fix their mistakes, they cover them up. Although they claim a 99 percent success rate for prosthetic knee surgery, there are far too many veterans who, like me, are unable to stand or walk after surgery for the number to be anywhere near that high. Surgeons should be doctoring patients rather than doctoring statistics.

Statistics are frequently fudged to make someone or something appear better than they actually are. Unfortunately, this practice is not limited to the local level. Top level VA administrators have for years touted the achievements of the VA healthcare system, however, a newly released VA study, "VA Health System Shines in Quality of Care Study," has been judged to be a prevarication by independent experts who have examined it. The research on which the study depends to reach its conclusions is out of date. None of it is less than five years old. One source is dated 1991, when George H.W. Bush was president. It is said that the study suffers from "publication bias," since much of the research cited was funded by the VA. Secretary Eric Shinseki says "the results speak for themselves." Most veterans advocates are saying the skewed results prove that the VA promotes itself at the expense of its patients. Veterans are insulted by the VA's attempt to pass off its promotional propaganda as a critical scientific analysis of healthcare data.

Shortly after taking office, President Obama issued a statement committing his Administration to maintaining "an unprecedented level of openness in Government" by orchestrating "a system of transparency, public participation, and collaboration."

His stated policy is admirable, but the reality is that the VA does not seem to have gotten the message. They continue to publish defective studies about what a good job VA healthcare is doing, rather than fixing the many defects in the system. It is ridiculous to claim that it shines in comparison to private healthcare. What happened to openness in government and transparency? Think of this as one more reason to cut back on the VA's dependence on its bloated Washington D.C. bureaucracy.

Since the VA is unwilling to provide me with a new artificial right knee, they should have to give me a voucher so that I can have the operation done by a responsible private practice surgeon in whom I have confidence. Having been painfully butchered once by a VA quack, I never want to go through that again. Congress gets excellent healthcare, why can't veterans?

Many veterans have already benefited from vouchers issued by the VA's Section 8 housing program. By using vouchers, the VA could supplement its substandard healthcare and give veterans the level of healthcare they deserve. Whenever bottlenecks are encountered, such as having to wait longer than five months for an appointment or a year for surgery, the VA should be required to issue vouchers that would permit veterans to get treated by a private physician.

It is time we ended the VA's monopoly on caring for veterans by introducing a modicum of competition from private doctors. This would also provide the VA with an incentive to do a better job of servicing veterans. I believe the way to motivate an intransigent bureaucracy is to light a fire under it. Anyway, it will get them up and moving in the right direction.

Immobile patients in VA nursing homes often die from infections stemming from bed sores. Prevention should be a top priority of nursing personnel. Bed bound patients have to be turned hourly in order to keep bed sores from developing. Yes, this consumes a lot of time and resources but we don't want to be buried in the VA cemetery any sooner than necessary. Those who must sit in a wheelchair for much of their waking hours should be given gel cushions to replace their foam eggcrate cushions. Water based ointments and medicated talcum powder can also prevent skin from chafing.

I am 62 years old, have been diagnosed with several disorders, yet I am not permitted to be seen by my primary physician more than once every six months for a half hour at most. Healthcare? I don't think so. I prefer to call it criminal neglect. With its limited resources, the VA compensates by extending the waiting period between appointments. Consequently, the doctor gets demoted to the status of a pill pusher and has no time for prevention due to his pressing case load. Reducing efforts at prevention results in a greater need for treatment. The whole process needs to be rethought.

During the Civil War, surgeons were derisively called

sawbones because they amputated limbs with bullet wounds. Their bodies were filthy and the leather aprons they wore around their waists were smeared in blood. Thousands of wounded soldiers died of infection following amputation of an arm or leg. Surprisingly, sterilization techniques aren't always adhered to by modern VA surgeons. Infections often take hold in patients following an operation. More emphasis needs to be placed on proper procedure. The surgical team must be careful to sterilize all instruments immediately after performing surgery. I am told that due diligence is what is most required to lower infection rates at VAMC facilities.

Why does the medical staff and administrators have broadband access to the internet, while patients don't? Having the ability to go online would empower inpatients to get second opinions and might diminish the authority and image of the medical staff. Will patients going online lead to distrust? Knowledge is dangerous, especially information that calls into question the steady stream of lies and propaganda that Veterans Affairs is trying to feed the public. Transparency and a corrupt bureaucracy cannot coexist. Sooner or later, one will displace the other.

Don't be afraid to complain about practices and procedures that affect you in a negative manner. Stand up for yourself and other veterans. The VA was created to help veterans; the people who administer it are supposed to serve us. Employees who can't or won't should be fired. It is our VA, not theirs.

Why not get rid of the VA's institutional-style nursing homes and replace them with assisted living facilities? Ailing veterans don't desire to lose their free will and independence. Don't treat them as if they are on their way out. We should be encouraging them to participate in social life, where they can serve as role models for children.

Why recruit medical personnel from Third World countries when we have record unemployment and cannot find jobs for our own citizens? It is high time for veterans to take charge and end this ridiculous nonsense. We need to take care of our own. The rest of the world is welcome to follow our lead. God helps those who help themselves.

When I was 19, the U.S. was at war. Uncle Sam drafted me out of college and taught me to kill. I learned my lessons well. Now, the enemy is the VA who first butchered me and then refused to put me back together. My instincts tell me to kill them all and let God sort them out. However, I am a Christian, and Jesus wants us to forgive and forget. But that only applies when the offender is sincerely repentant. For that to happen, the VA would have to undergo a complete transformation. That is our mission and I'm writing this book to assist us in accomplishing it. Be responsible and take charge. Veterans Affairs is our affair. Let's get moving.

Appendix A

2009 VA Co-pay Rates
(From VA Fact Sheet 16-1)

1. Outpatient Services
    A. The co-pay amount is limited to a single charge per visit.
    B. Services provided by a primary care clinician $15 / visit
    C. Services provided by a clinical specialist (includes MRI, CAT scan, and nuclear
        medicine studies $50 / visit
2. Medications
    A. For each 30-day or less supply of medication for treatment of non service-   connected condition $8
    B. Veterans in Priority Groups 2 through 6 are limited to a $960 annual cap
3. Inpatient Services
    A. Based on geographically-based means testing, lower income
        veterans who live in high-cost areas may qualify for a reduction of 80% of
        inpatient co-pay charges
    B. Inpatient co-pay for first 90 days of care during a 365-day period $1068
    C. Inpatient co-pay for each additional 90 days of care during a 365-day
        period $534

D. Per diem charge $10 / day

4. Long-term Care

A. Co-pays for long-term care start on the 22nd day of care during any 12-month

period; there is no co-pay requirement for the first 21 days; actual co-pay

charges will vary according to financial information submitted on VA Form 10-

10EC.

B. Nursing Home Care/Respite Care/Geriatric Evaluation   maximum of $97 / day

C. Adult Day Healthcare/Outpatient Geriatric Evaluation/Outpatient Respite Care

maximum of $15 / day

D. Domiciliary Care maximum of $5 / day

Appendix B

Facts Concerning the Department of Veterans Affairs

The Department of Veterans Affairs (VA) was established on March 15, 1989, a name change to upgrade the Veterans Administration which it had been called since the 1930's. The VA is the second-largest of the fifteen executive cabinet departments and operates nationwide programs for health care, education, financial assistance, and burial benefits.

Of the 23.4 million veterans currently alive, nearly three-quarters served during a war or an official period of conflict. Approximately a quarter of the nation's population is potentially eligible for VA benefits and services because they are veterans, family members or survivors of veterans.

The responsibility to care for veterans, survivors, spouses, and dependents can last a long time. Two children of Civil War veterans still draw VA benefits. About 184 children and widows of Spanish-American War veterans (circa 1898) still receive VA compensation and/or pensions.

The VA's fiscal year 2009 spending is projected to be about $93.4 billion, including $40 billion for health care, $46.9 billion for benefits, and $230 million for the national cemetery system. There is more than a seven percent increase from the agency's $87.6 billion budget for fiscal

year 2009.
Compensation and Pension

Disability compensation is a payment to veterans who are disabled by injury or disease incurred or aggravated during active military service. Wartime veterans with low incomes who are permanently and totally disabled may be eligible for financial support through the VA's pension program.

In fiscal year 2008, the VA provided $38.9 billion in disability compensation, death compensation or pension from the VA. In addition approximately 554,700 spouses, children and parents of deceased veterans received VA benefits. Among them are 170,144 survivors of Vietnam-era veterans and 235,000 survivors of World War II veterans.

Education and Training
Since 1944, when the first GI Bill began, more than 21.8 million veterans have received $83.6 billion in GI Bill benefits for education and training. The number of GI Bill recipients include 7.8 million veterans from World War II, 2.4 million from the Korean War and 8.2 million post-Korean and Vietnam-era veterans, plus active duty personnel.

Since the dependents program was enacted in 1956, the VA has also assisted in the education of more than 784,000 dependents of veterans whose deaths or total disabilities were service-connected. Since the Vietnam-era, there have been approximately 2.7 million veterans, service members, reservists and National Guardsmen who have participated in the Veterans' Educational Assistance Program (VEAP). VEAP was established in 1977and the Montgomery GI Bill in 1985.

In 2008, the VA helped pay for the education or training of 336,527 veterans and active-duty personnel, 106,092 reservists and National Guardsmen and 80,079 survivors.

Medical Care
Perhaps the most visible of all VA benefits and services is health care. From 54 hospitals in 1930, the VA's health care system now includes 153 medical centers, with at least one in each state, Puerto Rico and the District of Columbia. The VA operates more than 1,400 sites of care, including 909 ambulatory care and community-based outpatient clinics. 135 nursing homes, 47 residential rehabilitation treatment programs, 232 Veterans Centers and 108 comprehensive home care programs. VA health care facilities provide a broad spectrum of medical, surgical and rehabilitative care.

Almost 5.5 million people received care in VA health care facilities in 2008. By the end of fiscal year 2008, 78 percent of all disabled and low-income veterans had enrolled with VA for health care and community-based outpatient clinics; 65 percent of them were treated by the VA. In 2008 VA inpatient facilities treated 773,600 patients. The VA's outpatient clinics registered over 60 million visits.

The VA manages the largest medical education and health professions training program in the United States. VA facilities are affiliated with 107 medical schools, 55 dental schools and more than 1,200 other schools across the country. Each year, about 90,000 health professionals are trained in VA medical centers. More than half of the physicians practicing in the United States had some of their professional education in the VA health care system.

The VA's medical system serves as a backup to the Defense Department during national emergencies and as a federal support organization during major disasters.

In 1996, the VA put its health care facilities under 21 networks that provide more medical services to more veterans and family members than at any time during the VA's long history.

The VA has experienced unprecedented growth in the medical system workload over the past few years. The number of patients treated increased by 29 percent from 4.2 million in 2001 to nearly 5.5 million in 2008.

To receive VA health care benefits veterans must enroll. The VA health care system had nearly 7.9 million veterans who were enrolled as of October 2008. When they enroll, they are placed in priority groups or categories that help VA manage health care services within budgetary constraints and ensure quality care for those enrolled.

Some veterans are exempted from having to enroll. People who do not have to enroll include veterans with a service-connected disability of 50 percent or more, veterans who were discharged from the military within one year but have not been rated for a VA disability benefit and veterans seeking care for only a service-connected disability. Veterans with service-connected disabilities receive priority access to care for hospitalization and outpatient care. Veterans of Operation Enduring Freedom and Operation Iraqi Freedom (OEF/OIF) are eligible to receive enhanced health care benefits for five years following their military separation date.

Since 1979, the VA's Readjustment Counseling Service has operated Vet Centers, which provide psychological counseling for war-related trauma, community outreach and referral activities, plus supportive social services to veterans and family members. There are 232 Vet Centers.

Since the first Vet Center opened, more than two million veterans have been helped. Every year, the vet centers serve more than 130,000 veterans and accommodate more than a million visits by veterans and family members.

Vet centers are open to any veteran who served in the military in a combat theater during wartime or anywhere during a peiod of armed hostilities. Vet Centers also provide trauma counseling to veterans who were sexully assaulted or harassed while on active duty, and brereavement counseling to the families of service members who die on active duty.

The VA provides health care and benefits to more than 100,000 homeless veterans each year. Although the proportion of veterans among the homeless is declining, the VA continues to engage veterans in outreach, medical care, benefits assistance, transitional housing, and case management for veterans in permanent housing. The VA has made more than 450 grants for transitional housing, service centers and vans for outreach and transportation to state and local governments, tribal governments, non-profit community and faith-based service providers.

Indispensable to providing America's veterans with quality medical care are nearly 127,000 active volunteers in the VA's Voluntary Service who donated more than 11 million hours in 2008 to bring companionship and care to hospitalized veterans. These hours equate to 5,519 full-time employee-equivalent (FTEE) positions.

Research
VA research focuses on areas of concern to veterans. VA research has earned an international reputation for excellence in areas such as aging, chronic disease, prosthetics, and mental health. Studies conducted within the VA help improve medical care not only for the veterans enrolled in the VA's health care system, but for the nation at large. Because seven in ten VA researchers

are also clinicians, the VA is uniquely positioned to translate research results into improved patient care. VA scientists and clinicians collaborate across many disciplines, resulting in a synergistic flow of inquiry, discovery and innovation between labs and clinics.

VA investigators played key roles in developing the cardiac pacemaker, the CT scan, radioimmunoassay, and advances in artificial limbs. The first liver transplant in the world was performed by a VA surgeon-researcher. VA experiments on veterans established the effectiveness of new treatments for tuberculosis, schizophrenia and high blood pressure. The "Seattle Foot" developed by VA technicians has permitted veterans with amputations to run and jump. VA contributions to medical knowledge have won VA scientists many awards, including the Nobel Prize and the Lasker Award.

Special VA "centers of excellence" conduct leading-edge research in areas of prime importance to veterans, such as neurotrauma, prosthetics, spinal cord injury, hearing and vision Special VA "centers of excellence" conduct leading-edge research in areas of prime importance to veterans, such as neurotrauma, prosthetics, spinal cord injury, hearing and vision loss, alcoholism, stroke, and health care disparities. Through the VA's Cooperative Studies Program, researchers conduct multicenter clinical trials to investigate the best therapy for various diseases affecting large numbers of veterans. Examples of current projects include testing whether intensive control of blood sugar can reduce cardiovascular problems for patients with type 2 diabetes; and comparing deep brain stimulation with other treatments for Parkinson's disease.

Deployment health is a major priority for VA research. In addition to studies focused on recent veterans of operations Iraqi Freedom and Enduring Freedom, research continues on issues of special concern to veterans of earlier conflicts, such as the Gulf War and Vietnam War.

Home Loan Assistance From 1944, when VA began helping veterans purchase homes under the original GI Bill, through December 2007, more than 18.4 million VA home loan guaranties have been issued, with a total value of $967 billion. The VA ended fiscal year 2008 with almost 2.1 million active home loans, reflecting amortized loans totaling $220.8 billion.

In fiscal year 2007, the VA guaranteed 179,000 loans valued at $36.1 billion. During fiscal year 2008, the VA's programs for specially adapted housing helped 550 disabled veterans with grants totaling more than $24.6 million.

Insurance
The VA operates one of the largest life insurance programs in the world. VA directly administers six life insurance programs. In addition, the VA supervises the Servicemembers' Group Life Insurance and the Veterans' Group Life insurance programs. These programs provide $1.3 trillion in insurance coverage to 4 million veterans,

active-duty members, reservists and National Guard soldiers, plus 3.1 million spouses and children.

Traumatic Injury Protection program under Servicemembers' Group Life Insurance provides coverage to active-duty personnel who sustain traumatic brain injuries that result in severe losses. Benefit amounts range from $25,000 to $100,000, depending on the loss. This program covers 2.4 million members.

In 2007, the VA life insurance programs returned $354 million in dividends to 1 million veterans holding some of these VA life insurance policies, and paid an additional $1.1 billion in death claims.

Vocational Rehabilitation
The VA's Vocational Rehabilitation and Employment program provides services to enable veterans with service-connected disabilities to achieve optimum independence in daily living, and, to the maximum extent feasible, obtain and maintain employment. During fiscal years 1999 through 2008, 86,983 program participants achieved rehabilitation by obtaining and maintaining suitable employment. Additionally, during that same period 21,108 participants achieved rehabilitation through maximum independence in daily living.

The VA's National Cemeteries
In 1973, the Army transferred 82 national cemeteries to the VA, which now manages them through its National Cemetery Administration. Currently, the VA maintains 125

national cemeteries in 39 states and Puerto Rico.

In 2008, VA national cemeteries conducted 103,275 interments. The number is likely to increase as VA opens new national cemeteries or markers for veterans' graves. Since taking over the veterans cemetery program in 1973, the VA has provided more than 10.2 million headstones and markers.

Between 1999 and 2008, the VA opened 10 new national cemeteries. Gerald B. H. Solomon Saratoga National Cemetery near Albany, New York; Abraham Lincoln National Cemetery near Chicago; Dallas-Fort Worth National Cemetery; Ohio Western Reserve National Cemetery near Cleveland; Fort Sill National Cemetery near Oklahoma City; the National Cemetery of the Alleghenies near Pittsburgh; Great Lakes National Cemetery near Detroit; Georgia National Cemetery, north of Atlanta; Sacramento National Cemetery in California; and South Florida National Cemetery in West Palm Beach, Florida.

This year, the VA plans to open six new national cemeteries near Philadelphia; Jacksonville, Florida; Sarasota, Florida; Birmingham, Alabama; Greenville/Columbia, South Carolina; and Bakersfield, California. By 2009, these six new cemeteries will help the VA serve 90 percent of veterans with an open national cemetery or state veterans cemetery within 75 miles of their homes.

The VA administers the Presidential Memorial Certificate program, which provides gold embossed certificates to commemorate honorably discharged, deceased veterans. They are sent to the veteran's next of kin and loved ones. The VA provided 511,353 certificates in 2008.

The VA also administers the State Cemetery Grants Program, which encourages development of state veterans cemeteries. The VA provides up to 100 percent of the funds to develop, expand or improve veterans cemeteries operated by the states. More than $344 million has been awarded for 72 operational veterans cemeteries in 38 states, Saipan and Guam. In 2008, state cemeteries that received VA grants buried nearly 25,000 eligible and family members.

VA Employees
As of September 30, 2008, VA had 278,565 employees on the rolls. Among all departments and agencies of the federal government, only the Department of Defense has a larger workforce. Of the total number of VA employees, 247,113 were in the Veterans Health Administration, 16,135 in the Veterans Benefits Administration, 1,549 in the National Cemetery System, 3,412 in the Veterans Canteen Service and 437 in the Revolving Supply Fund. The rest, 9,919 employees, are in various staff and facilities offices.

Chronological History
In 1930, the Veterans Administration was created by Executive Order #5398, signed by President Herbert

Hoover on July 21. At that time, there were 54 hospitals, 4.7 million living veterans and 31,600 employees. In 1933, the Board of Veterans Appeals was established.

On June 22, 1944, a grateful President Franklin Roosevelt signed the "Serviceman's Readjustment Act of 1944" (Public Law 346, passed unanimously by the 78th Congress), more commonly known as "The GI Bill of Rights," offering home loans and education benefits to veterans.

The Department of Medicine and Surgery was also established, succeeded in 1989 by the Veterans Health Services and Research Administration, renamed the Veterans Health Administration in 1991.

In 1953, the Department of Veterans Benefits was established, succeeded in 1989 by the Veterans Benefits Administration.

In 1973, the National Cemetery System, renamed the National Cemetery Administration in 1998, was created when Congress transferred 82 national cemeteries from the Army to the VA. The Army kept Arlington National Cemetery and the U.S. Soldiers' and Airmens' Home National Cemetery in Washington, D.C.

In 1988, the legislation to elevate the VA to cabinet status was signed by President Ronald Reagan.

On March 15, 1989, the VA became the 14th department

in the presidential cabinet.

Secretaries of Veterans Affairs
Eric K. Shinseki	2009 - Present
James B. Peake	2007 - 2009
R. James Nicholson	2005 - 2007
Anthony J. Principi	2001 - 2005
Togo D. West, Jr.	1998 - 2000
Jesse Brown	1993 - 1997
Edward J. Derwinski	1989 - 1992

Produced courtesy of U.S. Department of Veterans Affairs - 810 Vermont Avenue, NW - Washington, DC 20420 Updated: November 10, 2009

Appendix C

July 08 FACT SHEET 06-03 Fraud, Waste and Abuse

What is health care fraud, waste and abuse?

Health care fraud and abuse occurs in every facet of the health care arena. Health care fraud is the intentional misrepresentation of a material fact on a health care claim in order to receive untitled payment. Health care waste and abuse describes practices that, either directly or indirectly, result in unnecessary costs to a health care program.

Some elements of fraud, waste and abuse may include:
• Misrepresentation or concealment of a material fact on a health care claim
• Knowledge that the facts on a medical claim is false or misrepresented
• Intent to deprive or harm the Health Administration Center (HAC) and its customers financially
• Unnecessary medical services or supplies
• Lack of conformity to professionally recognized standards
• Services or supplies rendered and billed at prices exceeding customary and usual charges

Who commits health care fraud, waste and abuse?

Providers who intentionally engage in any of the following are committing health care fraud, waste and abuse. This list is not all-inclusive:
• Bill incorrectly
• Bill for services you didn't get, inappropriate and

unnecessary services, including so-called "free services"
• Make false claims about qualifications, licensure and/or education
• Falsify records to suggest on-going medical services
• Forge a physician's signature on plans of care
• Alter information on care plans or prescriptions
• Unnecessary medical services or supplies
• Lack of conformity to recognized standards
• Services or supplies rendered and billed at prices exceeding customary and usual charges
• Falsify the diagnosis or procedure in order to maximize payments
• Change dates of service for double billing
• Waive the deductible and copays
Individuals who intentionally engage in any of the following commit health care fraud, waste and abuse. This list is not all-inclusive:
• Share health plan authorization cards
• Claim non-covered dependents
• Participate in doctor shopping ("Doctor Shopping" is a term commonly used to refer to a patient who may or may not have a legitimate physical ailment but goes from doctor to doctor with the objective of improperly obtaining multiple prescriptions for narcotic painkillers)
• Consent with providers to submit claims for services not received or not necessary
• Fabricate claims
• Alter submitted medical documentation of any type
• Use a stolen health plan authorization card to obtain health care services
• Use a deceased beneficiary's health plan authorization

card to obtain health care services
• Ineligible persons using a beneficiary's health plan authorization card to obtain medical services or benefits HAC employees who engage in any of the following acts commit health care fraud, waste and abuse. This list is not all-inclusive:
• Fabricate claims
• Provide false application data
• Change a provider's address to intercept provider payments
What are some things beneficiaries can do to assist in combating fraud, waste and abuse?
• Always protect your health plan authorization card
• Be cautious and know to whom you give your health plan authorization card or medical information to
• Immediately report a lost or stolen health plan authorization card
What should I do if I suspect fraud, waste or abuse?
Thoroughly review your Explanation of Benefits (EOB). If you note a service and/or supply billed to us that you did not receive, please report that immediately in writing. Please indicate in your letter that you are filing a fraud complaint and include the following facts:
• Name and address of the provider
• Name of beneficiary who was listed as receiving the service or item
• The claim number
• The date of the service in question
• The service or item that you do not believe was provided
• The reason and any supporting information or documentation why you believe the claim should not have

been paid

Whom should I contact if I suspect fraud, waste or abuse?

Please contact:
VA Health Administration Center
Attn: Program Integrity
PO Box 469060
Denver, CO 80246-9060
• Phone: 1-800-733-8387 Monday – Friday
• Fax: 1-303-331-7804
• Email: To contact us by email, please go to this web link and follow the directions for submitting secure email: http://www.va.gov/hac/contact
• Website: www.va.gov/hac

Appendix D

Survivors Benefits

What the VA Offers

The Department of Veterans Affairs (VA) offers a wide range of benefits and services for the surviving spouse, dependent children and dependent parents of deceased veterans and military service members.

Dependency and Indemnity Compensation (DIC) — DIC is a tax-free benefit for the surviving spouse and dependent children. A spouse's Survivor Benefits Plan (SBP) annuity is reduced by any DIC amount received. Note: A surviving spouse who remarries on or after December 16, 2003, and on or after attaining age 57, is entitled to continue to receive DIC.

VA also adds a transitional benefit of $250 to the surviving spouse's monthly DIC if there are children under age 18. The amount is based on a family unit, not individual children. It is paid for two years from the date that entitlement to DIC commences, but is discontinued earlier when there is no child under age 18 or no child on the surviving spouse's DIC for any reason.

To apply for DIC:
Usually an application for DIC benefits is completed by the Casualty Assistance Officer and submitted on behalf of the survivor. VA Form 21-534a, Application for

Dependency and Indemnity Compensation by a Surviving Spouse or Child, is used for this purpose. This form needs special processing and should be mailed or FAXED along with DD Form 1300, Report of Casualty, to:

Department of Veterans Affairs Regional Office and Insurance Center P.O. Box 8079 Philadelphia, PA 19101

FAX number: (215) 381-3084.

Parents' DIC -- Parents' DIC is a monthly benefit amount for the decedent's parents. It is based on income.

To apply for Parents DIC:
Download and complete a PDF version of VA Form 21-535 and mail to the VA regional office that serves the area where you reside.

Survivors' and Dependents' Educational Assistance (DEA) — Survivors' and Dependents' Educational Assistance provides payment of a monthly education or training allowance to the spouse and children of a veteran who died of a service-connected disability. Eligible persons can receive up to 45 months of benefits. Professional, educational and vocational counseling will be provided to eligible children and surviving spouses without charge upon request.

To apply for DEA:
Download and complete a PDF version of VA Form 22-5490. Follow the instructions on the form for more information on how to file the application for DEA benefits.

Work-Study Employment — This program is available to eligible survivors while pursuing a program of education or training under Dependent's Educational Assistance (Chapter 35).

To apply for Work Study Employment:
Download and complete a PDF version of VA Form 22-8691, Application for Work-Study Allowance. Follow the instructions on the form for more information on how to file the application for Work Study benefits.

Home Loan Guaranty — The surviving spouse of a veteran who died in service or as the result of a service-connected disability may be eligible for a guaranteed loan from a private lender. The loan may be used to purchase, construct or improve a home; to purchase a manufactured home and/or lot; or to refinance existing mortgages or other liens of record on a dwelling owned and occupied by the surviving spouse as his or her home. There is no time limit to use this benefit.

To apply for a VA Home Loan Guaranty:
Download and complete a PDF version of VA Form 26-1817, Request for Determination of Loan Guaranty Eligibility and submit it to the VA Loan Eligibility Center

that serves your location.

Burial Benefits (Headstones, Markers Presidential Memorial Certificates)
Headstones and Grave Markers VA provides headstones and grave markers for the graves of veterans anywhere in the world and of eligible dependents who are buried in military post, state veteran or national cemeteries. Niche markers also are available for identifying cremated remains in columbaria and memorial markers if the remains are not available for burial. Presidential Memorial CertificateA certificate bearing the President's signature is issued to recognize the service of deceased veterans who were discharged under honorable conditions. Eligible recipients include next of kin or other loved ones. A certificate can be issued to more than one eligible recipient. VA regional offices can help you in applying for certificates.

Life Insurance Settlement — Information on where and how to file for Servicemember's Group Life Insurance (SGLI) proceeds may be found at VA's Insurance Center Web site. You may also contact the Office of Service members' Group Life Insurance by phone at 1-800-419-1473, by email at osgli.claims@prudential.com, or by mail at:

Office of Servicemembers' Group Life Insurance 80 Livingston Avenue Roseland, New Jersey 07068-1733

Financial Counseling services are available at no cost to SGLI Insurance beneficiaries. This service provides a one-on-one counseling session, a detailed step-by-step financial plan, and access to financial counselors for one year. For additional information, call 1-888-243-7351.

Vet Center Bereavement Counseling — Bereavement Counseling is now being offered to parents, spouses and children of Armed Forces personnel who died in the service of their country. Also eligible are family members of reservists and National Guardsmen who die while on duty.

A new tri-fold brochure is now available for you to read or download. It is in a Power Point format but prints out nicely on two standard 8 1/2" by 11" sheets of paper. This is a 972 kb file.

Vocational Rehabilitation & Employment (VR&E) Services — VR&E can provide a wide range of vocational and educational counseling services to survivors and dependents who are eligible for one of VA's educational benefit programs. These services are designed to help an individual choose a vocational direction and determine the course needed to achieve the chosen goal.

Survivors and dependents should contact their local VA Vocational Rehabilitation & Employment program office for further information.

Education Program Refunds — The designated survivor of a deceased servicemenber will be refunded the service member's :
pay reductions for participation in the Montgomery GI Bill, less benefits previously paid to the servicemember contributions to the Veterans Educational Assistance Program (VEAP). VA Information and Assistance

Visit your VA regional office, or Call toll-free 1-800-827-1000, or Electronic Internet messaging at: https://iris.va.gov, or Visit the VA web site at http://www.va.gov. Information about State benefits may be found at: http://www.va.gov/partners/stateoffice/index.htm

Helpful Contacts

There are many veterans service organizations that offer assistance as well as other Federal and private organizations. For information on these organizations, refer to your telephone directory, contact your VA regional office, or visit the VA web site. Here is a list of the contacts available:

Direct Deposit (VA only) (877) 838-2778
VA Education Benefits (800) 442-4551
Service Member's Group Life Insurance (800) 419-1473
Benefits and Services Outside the U.S. (VA's Foreign

Services)

Telecommunication Device for the Deaf (TDD) (800) 829-4833

Social Security Administration (800) 772-1213

Survivor Benefit Plan (SBP) (800) 321-1080

TRICARE www.tricare.osd.mil (800) 874-2273

Office of Personnel Management (OPM) www.OPM.gov (800) 767-6738

American Gold Star Mothers www.goldstarmoms.com (202) 265-0991

Gold Star Wives of America www.goldstarwives.org (888) 751-6350

Tragedy Assistance Program for Survivors (TAPS) www.taps.org (800) 959-8277

Appendix E

Nursing Home Transparency and Improvement Act

The nursing home transparency provisions are the first comprehensive improvements in nursing home quality since the Omnibus Reconciliation Act of 1987. When the act is fully implemented, the law will provide consumers a substantial amount of new information about individual facilities, most of it from resolutions passed by members of the National Consumer Voice for Quality Long Term Care. The legislation was sponsored by Senator Herb Kohl (D-WI), Senator Chuck Grassley (R-IA), Rep. Henry Waxman (D-CA), Rep. Pete Stark (D-CA) and Rep. Jan Schakowsky (D-IL).

Key provisions of the act include:

Public disclosure of nursing home owners, operators, and other entities and individuals that provide management, financing, and services to nursing homes.

Establishment of internal procedures by nursing homes ("compliance and ethics programs") to reduce civil and criminal violations and improve quality assurance.

Collection of staffing data electronically from payroll records and other verifiable sources and public reporting of hours per resident day of care and turnover and retention rates.

Improved public information on Nursing Home Compare, including staffing data for each facility that includes hours of care per resident day, turnover, and retention rates; links to facilities' survey reports (Form 2567) and plans of correction on state websites; summaries of complaints against facilities, including number, type, severity and outcome; a standardized complaint form; and adjudicated criminal violations by facilities and their employees inside the facility, including civil monetary penalties levied against the facility, its employees, contractors, and other agents.

Establishment of a consumer rights information page on Nursing Home Compare, including services available from the long-term care ombudsman.

A review of Nursing Home Compare's accuracy, clarity, timeliness, and comprehensiveness and modifications of the site based on the review.

A Government Accountability Office study of the Five Star Quality Rating System.

Improved timeliness of survey information made available to the public.

A requirement for nursing homes to make surveys and complaint investigations for three years available on request and to post a notice that they are available.

A requirement that states maintain a website with

information on all nursing homes in the state, including survey reports (Form 2567), complaint investigation reports, plans of correction, and other information that the state or CMS considers useful.

A statutory requirement for a special focus facility program.

Establishment of a methodology for categorization and public reporting of facilities' expenditures, regardless of source of payment, for direct care (including nursing, therapy, and medical services); indirect care (including housekeeping and dietary services); capital assets; and administrative services. Improved complaint handling, including a voluntary standardized form for filing complaints with the survey agency and ombudsman; and protection of residents' legal representatives and other responsible parties from retaliation when they complain about quality of care.

Escrowing of civil monetary penalties after an independent informal dispute resolution process and pending resolution of further appeals. (Allows for reduction of CMP amounts for self-reported, non-repeat violations.)

Sixty-day advance notification of facility closure and authorization to continue Medicaid payments pending relocation of all residents.

Dementia care and abuse prevention in nurse aide

training programs.

Demonstration projects to identify best practices in culture change and information technology.

Demonstration program to develop, test, and implement federal oversight of interstate and large intrastate chains. (Chains apply to participate in the demonstrations.)

Appendix F

State Veteran's Nursing Homes

Get up to $1,949 a month from the Department of Veterans Affairs for veterans who served on active duty during World War II or the Korean Conflict or the Vietnam War. This extra income can be used to pay for home care or assisted living or nursing home care.

Get up to $1,056 a month from the VA for single widows or widowers of veterans who served on active duty during World War II or the Korean Conflict or the Vietnam War. This extra income can be used to pay for home care or assisted living or nursing home care.

Veterans Nursing Homes are generally available to active duty veterans but some states have beds for people who served with the reserves or National Guard and the spouses of veterans. The majority of these homes offer nursing care but some may offer assisted living or domiciliary care. Generally there is no income or asset test. Most veterans in most states would qualify. Many states have waiting lists of weeks to months for available beds. Each facility has different eligibility rules and there is an application process. You cannot simply walk in the door and arrange for nursing care on the spot. You must contact the veterans home you are interested in to find out the availability of beds and the application process.

There are other veteran benefits which make money available that may be a better solution to your care needs.

There are other veterans benefits that pay you money for private-pay nursing home care, assisted living, home care and other long term care needs.

There are also other services that may include free medical care, possible free prescription drugs, orthotics and prosthetics, home renovation grants for disabilities, home care, assisted living, domiciliary care, VHA nursing home care, and a possible host of other services or benefits for veterans. Qualifying veterans and spouses can receive a Pension Benefit income up to $1,949 a month for long term care assistance. To find out about these and how to apply, contact a Veterans Benefit Consultant in your area.

Alabama Bill Nichols State Veterans Home 1784 Elkahatchee Road Alexander City, AL 35010 (256) 329-3311
Alabama William F. Green State Veterans Home 300 Faulkner Drive Bay Minette, AL 36507-1461 (251) 937-8049

Alabama Floyd E. Tut Fann State Veterans Home 2701 Meridian Street Huntsville, AL 35811 (256) 851-2807
Alabama Pell City Pending

Alaska State Veterans And Pioneers Home 250 East Fireweed Palmer, AK 99645-6699 (907) 745-4241

The Arizona State Veteran Home 4141 North S. Herra Way Phoenix, AZ 85012 (602) 248-1550
Arizona State Veterans Nursing Home Tucson Pending

Arkansas Veterans Home 4701 West 20Th Street Little Rock, AR 72204 (501) 296-1885
Arkansas Fayetteville Veterans Home 1125 North College Fayetteville, AR 72703 (479) 444-7001

Veterans Home Of California, Barstow 100 East Veterans Parkway Barstow, CA 92311 (760) 252-6200
California Veterans Home Of California, Chula Vista 700 East Naples Court Chula Vista, CA 91911 (619) 482-6010
Veterans Home Of California, Yountville 100 California Drive Yountville, CA 94599 (707) 944-4500
Veterans Home of California - Lancaster 45221 30th Street West Lancaster, CA 93536 Pending
California Veterans Home of California - Fresno Pending
California Veterans Home of California - Ventura 10900 Telephone Road Ventura, CA 93004 Pending
California Veterans Home of California - Redding Pending

California West Los Angeles 11301 Wilshire Blvd Los Angeles, CA 90073 805-746-9115

Colorado State Veterans Home At Fitzimmons 1919 Quentin Street Aurora, CO 80010 (720) 857-6400
Homelake Colorado State Veterans Center 3749 Sherman Ave. Monte Vista, CO 81144 (719) 852-5118
Colorado Florence Colorado State Veterans Nursing Home 903 Moore Drive Florence, CO 81226 (719) 784-6331
Rifle Colorado State Veterans Nursing Home 851 East 5Th Street Rifle, CO 81650 (970) 625-0842
Walsenburg Colorado State Veterans Nursing Home 23500 U.S. Highway 160 Walsenburg, CO 81089 (719) 738-5133
Colorado Trinidad State Nursing Home 409 Benedicta Trinidad, CO 81082 (719) 846-9291

Connecticut Veteran's Home And Hospital 287 West Street Rocky Hill, CT 06067 (860) 721-5818
Delaware Veterans Home 100 Delaware Veterans Drive Milford, Delaware 19963 (302) 424-6000
Delaware Veterans Hospital, Nursing Home Care Unit 1601 Kirkwood Highway Wilmington, DE 19805 (302) 994-2511

Robert H. Jenkins, Jr. Veterans' Domiciliary Home Of Florida 751 S.E. Sycamore Terrace Lake City, FL 32025 (386) 758-0600
Florida Clifford Chester Sims State Veterans' Nursing 4419 Tram Road Springfield, FL 32404 (850) 747-5401

Florida Douglas Jacobson State Veterans' Nursing Home 21281 Grayton Terrace Port Charlotte, FL 33954 (941) 613-0919

Florida Emory L. Bennett Memorial 1920 Mason Avenue Daytona Beach, FL 32117 (904) 274-3460/61

Florida Baldomero Lopez State Veterans' Nursing Home 6919 Parkway Blvd Land-O-Lakes, FL 34639 (813) 558-5000

Florida Alexander Sandy Nininger State Veterans' Nursing Home 8401 West Cypress Drive Pembroke Pines, FL 33025 (954) 985-4824

Florida Clyde E. Lassen State Veterans Nursing Home 4650 State Road 16 St Augustine, FL 32092 N/A Scheduled to open year 2010

Georgia War Veterans Nursing Home 1101 15th Street Augusta, GA 30904 (706) 721-2824

Georgia War Veterans Home Carl Vinson Building Milledgeville, GA 1-888-453-6836

Hawaii Yukio Okutsu State Veterans Home 1180 Waianuenue Avenue Hilo, Hawaii 96720 (808) 961-1500

Idaho State Veterans Home - Boise 320 Collins Road Boise, ID 83702 (208) 334-5000

Idaho Lewiston Idaho State Veterans Home 821 21St Avenue Lewiston, ID 83501-6392 (208) 799-3422

Pocatello Idaho State Veterans Home 1957 Alvin Ricken

Drive Pocatello, ID 83201-2727 (208) 236-6340

Illinois Veterans Home Anna 792 N. Main Street Anna, IL 62906 (618) 833-6302
Illinois Veterans Home Lasalle 1015 O'connor Avenue Lasalle, IL 61301 (815) 223-0303 Ext. 200
Illinois Veterans Home Manteno 1 Veterans Drive Manteno, IL 60950 (815) 468-6581
Illinois Veterans Home Quincy 1707 N. 12Th Street Quincy, IL 62301 (217) 222-8641

Indiana Veterans Home 3851 N. River Rd West Lafayette, IN 47906-3762 (765) 463-1502

Iowa Veterans Home 1301 Summit St Marshalltown, IA 50158-5485 (641) 752-1501

Kansas Veterans' Home 1220 World War II Memorial Drive Winfield, KS 67156 (620) 221-9479 Ext. 250
Kansas Soldiers' Home 714 Sheridan Fort Dodge, KS 67801-9068 (620) 227-2121

Kentucky Thomson-Hood Veterans Center 100 Veterans Drive Wilmore, KY 40390 (859) 858-2814
Eastern Kentucky Veterans Center 200 Veterans Drive Hazard, KY 41701 (606) 435-6196
Western Kentucky Veterans Center 926 Veterans Drive Hanson, KY 42413 (270) 322-9087
Kentucky State Nursing Home Louisville Pending

Louisiana SE Louisiana War Veterans Home 4080 W Airline Hwy Reserve, LA 70084 (985) 479-4080
Louisiana War Veterans Home P.O. Box 748 Jackson, LA 70748-0748 (225) 634-5265
NE Louisiana War Veterans Home 6700 Highway 165 N. Monroe, LA 71203 (318) 362-4206
Louisiana NW Louisiana War Veterans Home 3130 Arthur Ray Teague Parkway Bossier City, LA 71112 (318) 741-2763
South West Louisiana War Veterans' Home 1610 Evangeline Road Jennings, LA 70546 (337) 824-2829

Maine Veterans' Home 1 Veterans Way RR #1 Box 11 Machias, ME 04654 (207) 255-0162
Maine Veterans 310 Cony Road Augusta, ME 04330 (207) 622-2454
Maine Veterans Home 44 Hogan Rd Bangor, ME 04401 (207) 942-2333
Maine Veterans Home 163 Van Buren Road Suite 2 Caribou, ME 04736 (207) 498-6074
Maine Veterans Home 290 U.S. Route 1 Scarborough, ME 04074 (207) 883-7184
Maine Veterans Home 477 High St. South Paris, ME 04281 (207) 743-6300

Maryland Charlotte Hall Veterans Home 29449 Charlotte Hall Road Charlotte Hall, MD 20622 (301) 884-8171

Massachusetts Soldiers' Home 110 Cherry Street Holyoke, MA 01041 (413) 532-9475 Ext. 126
Massachusetts Soldiers' Home 91 Crest Avenue Chelsea, MA 02150 (617) 884-5660 Ext. 210

Michigan Grand Rapids Home For Veterans 3000 Monroe, N. W. Grand Rapids, MI 49505 (616) 364-5400
Michigan D.J. Jacobetti Home For Veterans 425 Fisher St Marquette, MI 49855 (906) 226-3576

Minnesota Minneapolis Veterans Home 5101 Minnehaha Avenue South Minneapolis, MN 55417-1699 (612) 721-0600
Minnesota Hastings Veterans Home. 1200 E. 18Th St Hastings, MN 55033 (651) 438-8504
Minnesota Silver Bay Veterans Home 45 Banks Boulevard Silver Bay, MN 55614 (218) 226-6300
Minnesota Veterans Home P.O. Box 539 Luverne, MN 56156 (507) 283-1100
Minnesota Fergus Falls Veterans Home 1821 North Park Street Fergus Falls, MN 56537 (218) 736-0400

Mississippi Veterans Home 3466 Hwy. 80 East P.O. Box 5949 Pearl, MS 39288-5949 (601) 576-4850
Mississippi United States Naval Home (National) 1800 Beach Drive Gulfport, Mississippi 39507 (800) 322-3527
Mississippi State Veterans Home 3261 Hwy 49 Collins, MS 39428 (601) 765-0403

Mississippi State Veterans Home 4607 Lindberg Dr. Jackson, MS 39209 (601) 354-7205
Mississippi State Veterans Home 310 Autumn Ridge Dr. Kosciusko, MS 39090 (662) 289-7044
Mississippi State Veterans Home 120 Veterans Blvd. Oxford, MS 38655 (662) 236-7641

Missouri Veterans Home Mexico #1 Veterans Drive Mexico, MO 65265 (573) 581-1088
Missouri Veterans Home Cape Girardeau 2400 Veterans Memorial Drive Cape Girardeau, MO 63701 (573) 290-5870
Missouri Veterans Home St. Louis 10600 Lewis & Clark Blvd. St. Louis, MO 63136 (314) 340-6389
Missouri Veterans Home Cameron 1111 Euclid Cameron, MO 64429 (816) 632-6010
Missouri Veterans Home Warrensburg 1300 Veterans Road Warrensburg, MO 64093 (660) 543-5064
Missouri Veterans Home Mt. Vernon 1600 Hickory Mt. Vernon, MO 65712 (417) 466-7103
Missouri Veterans Home St. James 620 North Jefferson St. James, MO 65559 (573) 265-3271

Montana Veterans Home P.O. Box 250 Columbia Falls, MT 59912 (406) 892-3256
Montana Eastern Montana Veterans Home 2000 Montana Ave Glendive, MT 59330 (406) 377-8855

Nebraska Grand Island Veterans' Home 2300 Capital Ave. Grand Island, NE 68803-2097 (308) 385-6252
Nebraska Norfolk Veterans' Home 600 E. Benjamin Ave. Norfolk, NE 68701-0836 (402) 370-3177
Nebraska Thomas Fitzgerald Veterans' Home 15345 W. Maple Road Omaha, NE 68164-5186 (402) 595-2180
Western Nebraska Veterans' Home 1102 W. 42Nd St. Scottsbluff, NE 69361-4939 (308) 632-0300

Nevada Veterans' Nursing Home 100 Veterans Memorial Drive Boulder City, NV 89005 (702) 332-6864

New Hampshire New Hampshire Veterans Home 139 Winter Street Tilton, NH 03276 (603) 527-4400

New Jersey Veterans Memorial Home Vineland 524 North West Boulevard Vineland, NJ 08360-2895 (856) 696-6400
New Jersey Veterans Memorial Home Menlo Park 132 Evergreen Road Edison, NJ 18818-3013 (732) 452-4102
New Jersey Veterans Memorial Home Paramus 1 Veterans Drive Paramus, NJ 07653-0608 (201) 967-7676

New Mexico Veterans Home P.O. Box 927 Truth Or Consequences, NM 87901 (505) 894-4200
New Mexico Ft. Bayard Veterans Home P.O. Box 36219 Ft. Bayard, NM 88036 (505) 537-3302

New York New York State Veterans Home 220 Richmond Ave. Batavia, NY 14020 (716) 345-2000
New York State Veterans Home 4211 State Highway 220 Oxford, NY 13830-4305 (607) 843-3100
New York Long Island State Veterans Home 100 Patriots Road Stony Brook, NY 11790-3300 (631) 444-Vets
New York State Veterans Home 178-50 Linden Blvd Jamaica, NY 11437 (718) 481-6268
New York State Veterans Home 198 Albany Post Road Montrose, NY 10548 (914) 788-6000

North Carolina State Veteran Nursing Home 214 Cochran Ave. Fayetteville, NC 28301 (910) 482-4131
North Carolina State Veteran Home P O Box 599 Salisbury, NC 28145 (704) 638-4200
North Carolina North Carolina Veterans Home Kinston, NC Pending 2011
North Carolina North Carolina Veterans Home Asheville, NC Pending 2011

North Dakota Veterans Home 1400 Rose St. S. Lisbon, ND 58054 (701) 683-6501

Ohio Veterans Home 3416 Columbus Avenue Sandusky, OH 44870 1-800-572-7934
Southern Ohio Veterans Home 2003 Veterans Blvd Georgetown, OH 45121 (937) 378-2900

Oklahoma Lawton/Ft. Sill Veterans Center 501 S.E. Flowermound Road Lawton, OK 73501 (580) 351-6511
Oklahoma The Ardmore Veterans Center 1015 S. Commerce Ardmore, OK 73402 (580) 223-2266
Oklahoma The Claremore Veterans Center 3001 W. Blue Starr Drive Claremore, OK 74018-0988 (918) 342-5432
Oklahoma The Clinton Veterans Center Clinton, OK 73601 (580) 331-2200
Oklahoma Veterans Center 1776 East Robinson Street Norman, Oklahoma 73070 (405) 360-5600
Oklahoma Sulphur Veterans Center 200 E. Fairlane Sulphur, OK 73086 (580) 622-2144
Oklahoma Veterans Center Talihina, OK 74571 (918) 567-2251

Oregon Veterans' Home 700 Veterans' Drive The Dalles, OR (541) 296-7152

Pennsylvania Hollidaysburg Veterans Home P.O. Box 319 Hollidaysburg, PA 16648 (814) 696-5356 www.hvh.state.pa.us
Pennsylvania Soldiers' And Sailors' Home 560 East Third Street Erie, PA 165121 (814) 871-453 www.pssh.state.pa.us
Pennsylvania Southeastern Veterans Center Veterans Drive Spring City, PA 19475 (610) 948-2400 www.sevc.state.pa.us
Pennsylvania Gino J. Merli Veterans Center 401 Penn Avenue Scranton, PA 18503 (570) 961-4304 www.gmvc.state.pa.us
Pennsylvania Southwestern Veterans Center 7060

Highland Drive Pittsburgh, PA 15206 (412) 665-6708
www.swvc.state.pa.us
Pennsylvania Delaware Valley Veterans' Home 2701
Southampton Road Philadelphia, PA 19154-1205 (215)
965-5900 www.dvvh.state.pa.us

Puerto Rico Casa Del Veterano Carr 592,Km.5.6 Bo.
Amuelas Juan Diaz, PR 00795 (787) 837-6574

Rhode Island Rhode Island Veterans Home 480 Metacom
Ave. Bristol, RI 02809-0689 (401) 253-8000

South Carolina E. Roy Stone, Jr., Pavilion 2200 Harden
Street Columbia, SC 29203 (803) 737-5441
South Carolina Richard Michael Campbell Veterans
Nursing Home 4605 Belton Highway Anderson, SC
29621-7755 (803) 261-6734
South Carolina Veterans Victory House Nursing Home
2461 Sidneys Road Walterboro, SC 29488 843-538-3000

South Dakota Michael J. Fitzmaurice Veterans Home
2500 Minnekahta Ave. Hot Springs, SD 57747-1199 (605)
745-5127

Tennessee State Veterans Home 2865 Main Street
Humboldt, TN (731) 784-8405

Tennessee State Veterans Home 345 Compton Road Murfreesboro, TN 37130 (615) 895-8850
Tennessee State Veterans Home 9910 Coward Mill Road Knoxville, TN 37931-3101 (865) 862-8100
Tennessee State Veterans Home Montgomery County, TN Pending 2011

Texas William R. Courtney Texas State Veterans Home 1424 MLK, Jr. Lane Temple, TX 76504 (254) 791-8280
Frank M. Tejeda Texas State Veterans Home 200 Veterans Drive Floresville, TX 78114 (830) 216-9456
Ambrosio Guillen Texas State Veterans Home 9650 Kenworthy Street El Paso, Texas 79924 (915) 751-0967
Ussery-Roan Texas State Veterans Home 1020 Tascosa Road Amarillo, Texas 79124 (806) 322-8387
Alfredo Gonzalez Texas State Veterans Home McAllen, Texas 78503 (956) 682-4224
Lamun-Lusk-Sanchez Texas State Veterans Home 1809 North Highway 87 Big Spring, TX 79720 (432) 268-8387
Clyde W. Cosper Texas State Veterans Home 1300 Seven Oaks Road Bonham, TX 75418-3254 (903) 640-8387
Texas State Veterans Home Houston, TX Pending 2011

Utah Veterans Nursing Home 700 Foothill Drive Salt Lake City, UT 84113 (801) 584-1900
Utah George E. Wahlen Department of Veterans Affairs Medical Center 1102 N 1200 W Ogden, UT 84113 (801) 334-4300

Vermont Veterans Home 325 North Ave. Bennington, VT 05201 (802) 442-6353

Vermont Verdelle Village, Inc 596 Sheldon Road St. Albans, VT 05478 (802) 524-6534

Virginia Sitter & Barfoot Veterans Care Center 1601 Broad Rock Blvd Richmond, VA 23224 (804) 371-8433

Virginia Veterans Care Center 4550 Shenandoah Ave NW Roanoke, VA 24017 (540) 857-6974

Virginia State Veterans Home Hampton, VA Pending 2011

Washington Veterans Home 1141 Beach Dr. Retsil, WA 98378 (360) 895-4700

Washington Soldiers Home And Colony Orting-Kapowsin Hwy. Orting, WA 98360 (360) 893-4500

Washington Spokane Veterans' Home 225 E. 5Th Ave. Spokane, WA 99202 (509) 344-5770

Washington, D.C. United States Soldier's And Airmen's Home 3700 N. Capital Street N.W. Washington, D.C. 20317 (800) 422-9988

West Virginia West Virginia Veterans Home 512 Water St. Barboursville, WV 25504 (304) 736-1027

West Virginia State Veterans Home One Freedom Way

Clarksburg, WV 26301 (304) 626-1610

Wisconsin Veterans Home At King N2665 County Road King, WI 54946-0600 (715) 258-5586
Wisconsin Veterans Home At Union Grove 21425 Spring St. Union Grove, WI 53182 (262) 878-6700
Wisconsin State Veterans Home Chippewa Falls, WI Pending 2011

Veterans' Home Of Wyoming 700 Veterans Lane Buffalo, Wyoming 82834-9402 (307) 684-5511

Appendix G

Federal Nursing Homes Legislation

Older American Act , 1965

This Act created the Administration on Aging, authorized grants to states for aging-related community planning, services programs, research, demonstration and training projects; called for the development of State Units on Aging; and the "Aging Network," a web of federal, state, and local agencies linked together to focus on social services and other programs primarily targeted to older adults living in their homes.

Nursing Home Reform Act, 1987

In response to reports of widespread neglect and abuse in nursing homes in the 1980s, Congress enacted legislation in 1987 to require nursing homes participating in the Medicare and Medicaid programs to comply with certain requirements for quality of care. This law is included in the Omnibus Budget Reconciliation Act of 1987 (OBRA 1987), also known as the Nursing Home Reform Act. It specifies that a nursing home "must provide services and activities to attain or maintain the highest practicable physical, mental, and psychosocial well-being of each resident in accordance with a written plan of care..."

The Act was passed to ensure that nursing homes

residents' rights were maintained. It entitles all nursing home residents to receive quality care and live in an environment that maintains or improves the quality of their physical and mental health. This entitlement includes freedom from neglect, abuse, and misappropriation of property or funds. Neglect and abuse are criminal acts whether they occur inside or outside a nursing home. Residents do not surrender their rights to protection from criminal acts when they enter a facility.

To participate in the Medicare and Medicaid programs, nursing homes must be in compliance with the federal requirements for long term care facilities as prescribed in the U.S. Code of Federal Regulations (42 CFR Part 483).

Residents' Bill of Rights

The right to freedom from abuse, mistreatment, and neglect

The right to freedom from physical restraints

The right to privacy

The right to accommodation of medical, physical, psychological, and social needs

The right to participate in resident and family groups

The right to be treated with dignity

The right to exercise self-determination

The right to communicate freely

Required Services

Under the regulations, the nursing home must:

Have sufficient nursing staff.

Conduct initially a comprehensive and accurate assessment of each resident's functional capacity.

Develop a comprehensive care plan for each resident.

Prevent the deterioration of a resident's ability to bathe, dress, groom, transfer and ambulate, toilet, eat, and to communicate.

Provide, if a resident is unable to carry out activities of daily living, the necessary services to maintain good nutrition, grooming, and personal oral hygiene.

Ensure that residents receive proper treatment and assistive devices to maintain vision and hearing abilities.

Ensure that residents do not develop pressure sores and, if a resident has pressure sores, provide the necessary treatment and services to promote healing, prevent infection, and prevent new sores from developing.

Provide appropriate treatment and services to incontinent residents to restore as much normal bladder functioning as possible.

Ensure that the resident receives adequate supervision and assistive devices to prevent accidents.

Maintain acceptable parameters of nutritional status.

Provide each resident with sufficient fluid intake to maintain proper hydration and health.

Ensure that residents are free of any significant medication errors.

Promote each resident's quality of life.

Maintain dignity and respect of each resident.

Ensure that the resident has the right to choose activities, schedules, and health care.

Provide pharmaceutical services to meet the needs of each resident.

Be administered in a manner that enables it [the nursing home] to use its resources effectively and efficiently.

Maintain accurate, complete, and easily accessible clinical records on each resident.

Appendix H

Homeless Veterans

BACKGROUND

Far too many veterans are homeless in America—between 130,000 and 200,000 on any given night—representing between one fourth and one-fifth of all homeless people. Three times that many veterans are struggling with excessive rent burdens and thus at increased risk of homelessness. Further, there is concern about the future. Women veterans and those with disabilities including post traumatic stress disorder and traumatic brain injury are more likely to become homeless, and a higher percentage of veterans returning from the current conflicts in Afghanistan and Iraq have these characteristics.

The U.S. Department of Veterans Affairs estimates that 131,000 veterans are homeless on any given night. And approximately twice that many experience homelessness over the course of a year. Conservatively, one out of every three homeless men who is sleeping in a doorway, alley or box in our cities and rural communities has put on a uniform and served this country.

Approximately 40% of homeless men are veterans, although veterans comprise only 34% of the general adult male population. The National Coalition for Homeless Veterans estimates that on any given night, 200,000 veterans are homeless, and 400,000 veterans will experience homelessness during the course of a year (National Coalition for Homeless Veterans, 2006). 97% of those homeless veterans will be male (Department of Veterans Affairs, 2008).

DEMOGRAPHICS

The U.S. Department of Veterans Affairs (VA) says the nation's homeless veterans are mostly males (four percent are females). The vast majority is single, most come from poor, disadvantaged communities, 45 percent suffer from mental illness, and half have substance abuse problems. America's homeless veterans have served in World War II, Korean War, Cold War, Vietnam War, Grenada, Panama, Lebanon, Operation Enduring Freedom (Afghanistan), Operation Iraqi Freedom, or the military's anti-drug cultivation efforts in South America. 47 per cent of homeless veterans served during the Vietnam Era. More than 67 per cent served our country for at least three years and 33 per cent were stationed in a war zone.

Here are some statistics concerning the veterans homeless:

23% of homeless population are veterans
33% of male homeless population are veterans
47% Vietnam Era
17% post-Vietnam
15% pre-Vietnam
67% served three or more years

33% stationed in war zone
25% have used VA Homeless Services
85% completed high school/GED, compared to 56% of non-veterans
89% received Honorable Discharge
79% reside in central cities
16% reside in suburban areas
5% reside in rural areas
76% experience alcohol, drug, or mental health problems
46% white males compared to 34% non-veterans
46% age 45 or older compared to 20% non-veterans

Female homeless veterans represent an estimated 3% of homeless veterans. They are more likely than male homeless veterans to be married and to suffer serious psychiatric illness, but less likely to be employed and to suffer from addiction disorders. Comparisons of homeless female veterans and other homeless women have found no differences in rates of mental illness or addictions.

PROGRAMS AND POLICY ISSUES

While most housing help available to veterans focuses on homeownership, there have been Federal investments in programs for homeless veterans. The Department of Veterans' Affairs (VA) funds temporary housing for homeless veterans including:

• shelter and two-year transitional housing funded through the Grant and Per Diem Program,
• long-term care through the Domiciliary Care for

Homeless Veterans Program, and
• skills programs such as the Compensated Work Therapy/Veterans Industries Program.

These programs do not meet existing need. For example, Grant and Per Diem only funds 8,000 beds. In addition, the Department of Housing and Urban Development (HUD) works with VA to operate the HUD-VA Supportive Housing (HUD-VASH) program. HUD-VASH connects HUD Housing Choice Vouchers with VA case management and services. This is HUD's only program targeted directly to veterans. HUDVASH, a long standing and rigorously tested program, has been under-resourced in past years, but the recent addition of 10,000 vouchers a year for two years has been a crucial step forward. The Administration did not request additional vouchers for 2010. However, the program is popular in Congress, and there is a strong possibility of additional vouchers this year.

VA's Homeless Providers Grant and Per Diem Program

The Grant and Per Diem program is offered annually (as funding permits) by the VA to fund community-based agencies (up to 65% of a given project) providing transitional housing or service centers for homeless veterans.

While most housing help available to veterans focuses on homeownership, there have been Federal investments in programs for homeless veterans. The Department of

Veterans' Affairs (VA) funds temporary housing for homeless veterans including:

• Shelter and two-year transitional housing funded through the grant and per Diem Program,
• Long-term care through the Domiciliary Care for Homeless Veterans Program, and
• Skills programs such as the Compensated Work Therapy/Veterans Industries Program.

These programs do not meet existing need. For example, Grant and Per Diem only funds 8,000 beds.

In addition, the Department of Housing and Urban Development (HUD) works with VA to operate the HUD-VA Supportive Housing (HUD-VASH) program. HUD-VASH connects HUD Housing Choice Vouchers with VA case management and services.

In VA's Compensated Work Therapy/Transitional Residence (CWT/TR) Program, disadvantaged, at-risk, and homeless veterans live in supervised group homes while working for pay in VA's Compensated Work Therapy Program (also known as Veterans Industries). Veterans in the CWT/TR program work about 33 hours per week, with approximate earnings of $732 per month, and pay an average of $186 per month toward maintenance and up-keep of the residence. The average length of stay is about 174 days. VA contracts with private industry and the public sector for work done by these veterans, who learn new job skills, relearn successful work habits, and regain a sense of self-esteem and self-worth.

Supported Housing

In 2008, according to the annual homeless assessment report to Congress, 3% of the shelter's beds were reserved for the veterans. Like the HUD-VASH program, staff in VA's Supported Housing Program provides ongoing case management services to homeless veterans. Emphasis is placed on helping veterans find permanent housing and providing clinical support needed to keep veterans in permanent housing. Staff in these programs operate without benefit of the specially dedicated Section 8 housing vouchers available in the HUD-VASH program but are often successful in locating transitional or permanent housing through local means, especially by collaborating with Veterans Service Organizations.

In addition, the VA extends loans, funds Veterans Benefits Counselors, and operates drop-in centers where veterans can clean up and receive therapeutic treatment during the day.

The National Coalition for Homeless Veterans estimates that the VA serves about 25% of veterans in need – a figure that would leave approximately 300,000 veterans each year to seek assistance from local government agencies and voluntary organizations.

In general, the needs of homeless veterans do not differ

from those of other homeless people. The National Coalition for Homeless Veterans suggests the most effective programs are "community-based, nonprofit, 'veterans helping veterans' groups" (NCHV "Background and Statistics"). However there is some evidence that programs which recognize and acknowledge veteran experience may be more successful in helping homeless veterans transition into stable housing. Until serious efforts are made to address the underlying causes of homelessness, including inadequate wages, lack of affordable housing, and lack of accessible, affordable health care, the tragedy of homelessness among both veterans and non-veterans will continue to plague American communities.

Bibliography

Adams, Chris, "Data Questioned: VA Touts Care Study, Though Issues Linger," The Press-Enterprise (Riverside, CA), Nov 13, 2010, pp. A1, A11.

Doyle, Sue, "Vets' Backs to the Wall Protest: Federal Decision Takes Land Away from Home for Former Military Members," Daily News (Los Angeles), March 10, 2008.

Pearlman, Robert, Helene Starks, Kevin Cane, William Cole, David Rosengren, and Donald Patrick, Your Life, Your Choices, Brems Eastman & Partners, Seattle, 53 pp.

Phelps, M. William, The Devil's Rooming House, Lyons Press, 2010, Guilford, CT, 304 pp.

Pool, Bob, "Veterans with a Gripe Upend Stars and Stripes," Los Angeles Times, June 27, 2009, A8.

Acknowledgements

Nazi Germany perfected the art of record keeping. Almost every action required filling in one or more specialized forms. Meticulous lists were kept of the millions of Roma, homosexuals, Jews, and other undesirables that had been gassed with Cyclon B and cremated in the Third Reich's concentration camps.

These records provided evidence for the Nuremberg War Crimes trials immediately following World War II. Justice was ultimately done with the assistance of the Germans' own records.

The Veterans Administration has gone even further by digitalizing the medical files of millions of veterans, some of whom suffered greatly by being used as guinea pigs in an ongoing program of medical experimentation. Most were volunteers only in the sense that the VA volunteered them. Malpractice, abuse, and negligence; it has all been recorded in black and white for future presentation in a court of law.

During my two year stint at Loma Linda VA nursing homes, record keeping was done by means of ordinary Windows personal computers. Physicians were able to download porn, chat with underage girls and gamble online during working hours without much danger of being caught.

They are much more efficient now. Recently, the VA

purchased brand new Dell computers in volume and connected them to an in-house administrative network which records keystrokes and websites visited. Quacks getting their jollies at XXX websites and arranging to take part in bacchanalian orgies is, thank goodness, largely a thing of the past.

Special thanks to all of the doctors who didn't bother to log off their computers at night. The lax security permitted me to upload files to my website and document many of the medical mistakes that I saw take place. As I said earlier, it doesn't do any good to deny what you typed on the keyboard. But don't worry, murder as many of us as you like, quacks rarely go to prison. The Golden Scalpel Award goes to head of orthopedic surgery, Dr. Barton, who absolutely refuses to acknowledge and subsequently correct his team's surgical errors. No doubt he was born to butcher.

Veterans Affairs has its own police force. For the most part, the officers maintain security, issue parking tickets, and assist nursing home residents with physical and mental problems. When I explored the hallways at night in my wheelchair, I was grateful for their presence. If I got tired, I could count on them to push me back to the nursing home. In return I downloaded video games off the internet such as *Bejeweled* and *Solitaire* for them to play during their coffee breaks. Once, when I got in a fight with another veteran in the pharmacy, two officers pulled us apart with a minimum of force and kept us away from each other until we cooled off. What impressed me the most was that they evidently got to the root of the problem. The next time I went to the pharmacy, the wait had been reduced from four hours to less than 40 minutes. Also, a video monitor listing the names of the veterans whose prescriptions had been filled was strategically placed in the Food Court on the Second Floor where patients could relax and have something to eat while waiting for their medications.

Last, but not least, I would like to thank the volunteers who provide escort service and staff the information desk. Since they are not compensated for their services, they must do what they do out of the goodness of their hearts.

www.ingramcontent.com/pod-product-compliance
Lightning Source LLC
Chambersburg PA
CBHW030006190526
45157CB00014B/557